Collaborative Development for the Prevention of Occupational Accidents and Diseases

Rodolfo Andrade de Gouveia Vilela
Marco Antonio Pereira Querol
Sandra Lorena Beltran Hurtado
Gislaine Cecília de Oliveira Cerveny
Manoela Gomes Reis Lopes
Editors

Collaborative Development for the Prevention of Occupational Accidents and Diseases

Change Laboratory in Workers' Health

Springer

Editors
Rodolfo Andrade de Gouveia Vilela
Department of Environmental Health
School of Public Health
University of São Paulo
São Paulo, SP, Brazil

Sandra Lorena Beltran Hurtado
School of Public Health
University of São Paulo
São Paulo, SP, Brazil

Manoela Gomes Reis Lopes
Department of Biological Sciences
and Health
Federal University of Amapá
Macapá, AP, Brazil

Marco Antonio Pereira Querol
Department of Agronomic Engineering
Federal University of Sergipe
São Cristóvão, SE, Brazil

Gislaine Cecília de Oliveira Cerveny
School of Public Health
University of São Paulo
São Paulo, SP, Brazil

ISBN 978-3-030-24419-4 ISBN 978-3-030-24420-0 (eBook)
https://doi.org/10.1007/978-3-030-24420-0

This Springer imprint is published by the registered company Springer Nature Switzerland AG
The registered company address is: Gewerbestrasse 11, 6330 Cham, Switzerland

In opposition to the darkness and obscurantism that hang over the planet and in memory of Paulo Freire (1921–1997) we dedicate this book to all the researchers and professors that take sides for critical thinking and human dignity

(...) There is no such thing as teaching without research and research without teaching. One inhabits the body of the other (...) I teach because I search, because I question, and because I submit myself to questioning. I research because I notice things, take cognizance of them. And in so doing, I intervene. And intervening, I educate and educate myself (...)

Freire P. Pedagogy of Freedom: Ethics, Democracy, and Civic Courage. Rowman & Littlefield Publishers, 1998

Foreword

Change Laboratories Facing the Future

There is a growing need across the social sciences to explore new methodologies and methods of interventionist research. The methodology of formative interventions, implemented in the form of Change Laboratories, "represents perhaps the most visible program of contemporary intervention research" (Penuel et al. 2016, p. 489). Scaratti and his coauthors (2017, p. 62) point out that Change Laboratories offer an example of "how a robust theory can enhance and shape an intervention methodology in order to develop the building of dynamic possibility knowledge." Greeno concludes that further analyses of Change Laboratory interventions "would indeed provide a significant advance in the field's understanding of learning and development" (Greeno 2016, p. 637).

The Change Laboratory is a formative intervention method based on cultural-historical activity theory and the theory of expansive learning (Engeström et al. 2014; Sannino et al. 2016; Sannino and Engeström 2017; Virkkunen and Newnham 2013). Committed to serving struggles for equity, social justice, and sustainability, the Change Laboratory was developed and first implemented in 1995 in Finland. Grounded in materialist dialectics and the Vygotskian theoretical tradition, its uptake has been demanding and only quite recently it has reached the point of accelerated international adoption. By now Change Laboratories have been used in some 30 countries in educational institutions, workplaces, and communities, to generate bottom-up solutions to challenging problems and complex transformations.

The Change Laboratory offers four important contributions to social sciences. First, it bridges bold practical change efforts with a rigorous activity–theoretical framework (*the theory–practice contribution*; Sannino and Engeström 2017). Second, it makes accessible and analyzable the emergence and formation of the participants' transformative agency, understood as grappling with contradictions and conflicts of motives by means of double stimulation (*the agency contribution*; Sannino et al. 2016). Third, the Change Laboratory reveals the potential and conditions of learning by ascending from the abstract to the concrete, understood as construction and expansive use of novel theoretical concepts (*the expansive learning contribution*; Sannino et al. 2016). Fourth, the Change Laboratory offers a novel methodology and toolkit of methods of data analysis for gaining insight into complex processes of transformation (*the research instrument contribution*; Engeström and Sannino 2019).

Aimed at expansive learning, Change Laboratory interventions put into use the key principles of double stimulation (Sannino 2015) and ascending from the abstract to the concrete (Engeström and Sannino 2016), typically with a carefully planned sequence of "mirror materials," tasks and learning actions. This does not mean that the interventionist's plan is smoothly implemented. To the contrary, participants in formative interventions commonly take over the process at some point and generate deviations from the interventionist's intentions. These deviations reveal gaps between the interventionist's object and the participants' objects – gaps that need to be negotiated. The deviations, gaps, and negotiations are important instances of emerging transformative agency among the participants (Engeström and Sannino 2012; Engeström et al. 2013).

The actual sessions of a Change Laboratory intervention need to be seen in the context of a broader formative research design. This may be represented with the help of Fig. 1.

In Fig. 1, research moves bidirectionally, from bottom up and from top down. The analysis of focal data is the meeting point where theory and data come together

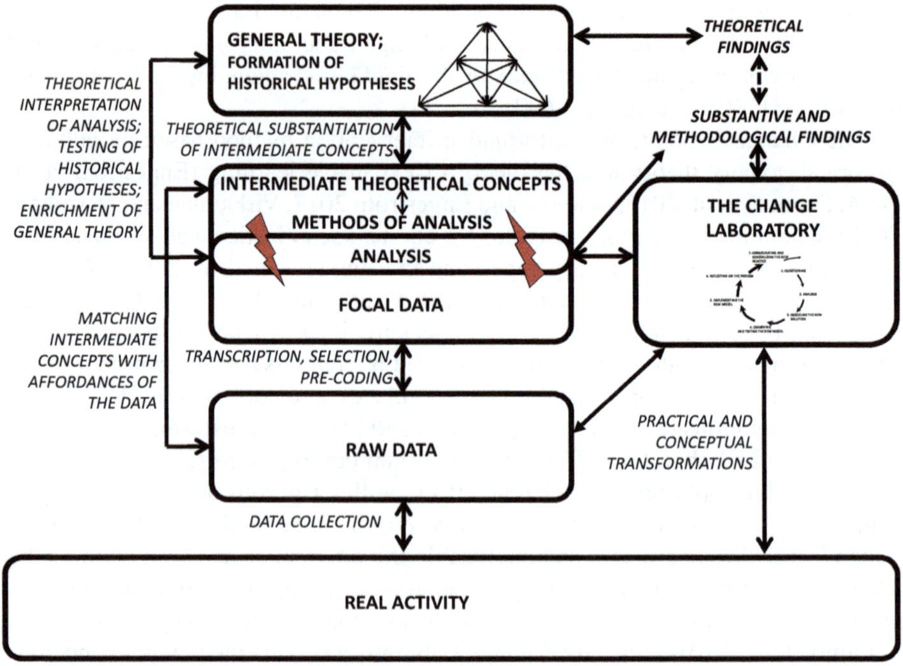

Fig. 1 General design of formative intervention research

and collide, as indicated by means of the lightning images in the figure. The collision is a source of new theoretical and methodological developments.

The box on the top of Fig. 1 (General theory) signifies the formulation of a historically grounded working hypothesis concerning the contradictions of the activity system (or activity systems) under investigation. The triangular icon within this box indicates that the working hypothesis is generated with the help of the conceptual lens of the general model of an activity system (Engeström 2015, p. 63). The movement of research has also a sideways dimension. In Fig. 1, the Change Laboratory is depicted horizontally adjacent and reciprocally connected to the analysis of focal data. This means on the one hand that parts of the analysis are collaboratively conducted in the Change Laboratory sessions, and, on the other hand, that instruments and outcomes of data analysis are used in design efforts in the Change Laboratory. The cycle-shaped icon in the Change Laboratory box indicates that the intervention facilitates the stepwise accomplishment of an expansive learning cycle (Engeström and Sannino 2010).

The inclusion of the Change Laboratory changes the dynamics of the research design. Practical and conceptual transformations generated in the intervention are implemented in real activity, generating new data that is fed into the analysis. The research process takes on a longitudinal and cyclic character, requiring multiple iterations of data collection, analysis, and theoretical interpretation. This why the approach is also called "developmental work research" (Engeström 2005).

Observations made by authors of the present book (summarized in Chap. 15) indicate important challenges that have been encountered in Change Laboratory interventions conducted thus far. Interventions often need to be negotiated and aligned with already ongoing "intraventions," change efforts initiated from within (Engeström et al. 2016). Sometimes interventions have stopped before reaching the phases of implementation and consolidation of the created new models, which is obviously a challenge to the durability and sustainability of the new solutions. Also, Change Laboratory interventions have often been conducted in a single bounded activity system or a pair of closely interconnected activity systems. This may lead to encapsulation and consequently to diminished impact (Bal et al. 2018, p. 1044). In some Change Laboratories, the local bottom-up change effort has led to solutions that clashed with the practices, priorities, and power constellations of their larger organizational contexts, resulting in stalled implementation. This phenomenon, pointed out by Kerosuo (2006), was discussed by Blackler (2009, p. 37). He suggested that "the theory of expansive learning needs to be extended to help people better anticipate and manage the complex dynamics of organizational change"; thus "activity-theoretical studies of senior management teams would be of considerable interest in this respect." Taylor (2009, p. 238) echoed this observation, stating that we need to study "how the activity re-creates the larger system of activity".

The authors of the present book make it clear that Change Laboratories need to tackle these challenges. In addition to single cycles of expansion, multiple successive, parallel, and intertwined cycles need to be reconstructed, fostered, and analyzed. Formative interventions need to be increasingly distributed in multiple interconnected sites in a network or coalition of activities. Formative interventions need to involve actors and decision-makers from different levels or layers of such fields of activity. The resulting perspective is formative intervention research in heterogenous work coalitions (Sannino 2017).

This methodological perspective corresponds to the current efforts of articulating a fourth generation of cultural-historical activity theory. The focus of the fourth generation is critical junctures of societal transformation toward sustainability and equity. The fourth generation engages in enacting utopias or alternatives to capitalism. In Chap. 15 of this book, the authors ask "if the Change Laboratory has a future in a context of savage capitalism that is configured in Brazil and other continents." The authors answer their own question by pointing out that capitalism is essentially contradictory, unstable, and continuously changing. Thus, the future "needs to be cultivated within the old … showing its potential to answer questions that the old cannot answer."

The authors of this book, together with the diverse participants of their Change Laboratory interventions, demonstrate that formative intervention research can become an expanding collective force that addresses burning societal problems, cutting across disciplinary, institutional, and cultural boundaries without giving up methodological rigor. Similar dynamics are developing in other locations, notably in South Africa, in the research group of Heila Lotz-Sisitka and her colleagues (2017). The present book may be read as a set of interconnected spearheads moving into the zone of proximal development that we call fourth-generation activity theory and its methodology of formative interventions.

March 2019
Center for Research on Activity,
Development and Learning – CRADLE,
University of Helsinki,
Helsinki, Finland

Yrjö Engeström

References

Bal, A., Afacan, K., & Cakir, H. I. (2018). Culturally responsive school discipline: Implementing Learning Lab at a high school for systemic transformation. *American Educational Research Journal, 55*(5), 1007–1050. https://doi.org/10.3102/0002831218768796.

Blackler, F. (2009). Cultural-historical activity theory and organization studies. In A. Sannino, H. Daniels, & K. D. Gutiérrez (Eds.), *Learning and expanding with activity theory* (pp. 19–39). Cambridge: Cambridge University Press.

Engeström, Y. (2005). *Developmental work research: Expanding activity theory in practice*. Berlin: Lehmanns Media.

Engeström, Y. (2015). *Learning by expanding: An activity-theoretical approach to developmental research* (2nd ed.). Cambridge: Cambridge University Press.

Engeström, Y., Rantavuori, J., & Kerosuo, H. (2013). Expansive learning in a library: Actions, cycles and deviations from instructional intentions. *Vocations and Learning, 6*(1), 81–106. https://doi.org/10.1007/s12186-012-9089-6.

Engeström, Y., & Sannino, A. (2010). Studies of expansive learning: Foundations, findings and future challenges. *Educational Research Review, 5*(1), 1–24. https://doi.org/10.1016/j.edurev.2009.12.002.

Engeström, Y., & Sannino, A. (2012). Whatever happened to process theories of learning? *Learning, Culture and Social Interaction, 1*(1), 45–56. https://doi.org/10.1016/j.lcsi.2012.03.002.

Engeström, Y., & Sannino, A. (2016). Expansive learning on the move: Insights from ongoing research/El aprendizaje expansivo en movimiento: aportaciones de la investigación en curso. *Infancia y Aprendizaje., 39*(3), 401–435. https://doi.org/10.1080/02103702.2016.1189119.

Engeström, Y., & Sannino, A. The Change Laboratory as a research instrument: from methodology to methods and back. In *35th EGOS Colloquium*. Edinburgh, Scotland. Jul 4 –Jul 6, Forthcoming 2019.

Engeström, Y., Sannino, A., & Virkkunen, J. (2014). On the methodological demands of formative interventions. *Mind, Culture, and Activity, 21*(2), 118–128. https://doi.org/10.1080/10749039.2014.891868.

Greeno, J. G. (2016). Cultural-historical activity theory/design-based research in Pasteur's quadrant. *The Journal of the Learning Sciences, 25*(4), 634–639. https://doi.org/10.1080/10508406.2016.1221718.

Kerosuo, H. (2006). *Boundaries in action: an activity-theoretical study of development, learning and change in health care for patients with multiple and chronic illnesses*. Helsinki: Faculty of Behavioural Sciences, University of Helsinki.

Lotz-Sisitka, H., Mukute, M., Chikunda, C., Baloi, A., & Pesanayi, T. (2017). Transgressing the norm: transformative agency in community-based learning for sustainability in southern African contexts. *International Review of Education, 63*(6), 897–914. https://doi.org/10.1007/s11159-017-9689-3.

Penuel, W. R., Cole, M., & O'Neill, D. K. (2016). Introduction to the special issue. *The Journal of the Learning Sciences, 25*(4), 487–496. https://doi.org/10.1080/10508406.2016.1215753.

Sannino, A. (2015). The principle of double stimulation: a path to volitional action. *Learning, Culture and Social Interaction, 6*, 1–15. https://doi.org/10.1016/j.lcsi.2015.01.001.

Sannino, A. (2017). Researching work and learning for enacted utopias: the struggle to overcome homelessness. (Keynote Speaker). In *10th International Conference on Researching Work and Learning; 2017 Dec 7th – 8th*. Grahamstown: Rhodes University, South Africa.

Sannino, A., & Engeström, Y. (2017). Co-generation of societally impactful knowledge in Change Laboratories. *Management Learning, 48*(1), 80–96. https://doi.org/10.1177/1350507616671285.

Sannino, A., Engeström, Y., & Lemos, M. (2016). Formative interventions for expansive learning and transformative agency. *The Journal of the Learning Sciences, 25*(4), 599–633. https://doi.org/10.1080/10508406.2016.1204547.

Scaratti, G., Galuppo, L., Gorli, M., Gozzoli, C., & Ripamonti, S. (2017). The social relevance and social impact of knowledge and knowing. *Management Learning, 48*(1), 57–64. https://doi.org/10.1177/1350507616680563.

Taylor, J. R. (2009). The communicative construction of community: authority and organizing. In A. Sannino, H. Daniels, & K. D. Gutiérrez (Eds.), *Learning and expanding with activity theory* (pp. 228–239). Cambridge: Cambridge University Press.

Virkkunen, J., & Newnham, D. S. (2013). *The change laboratory: a tool for collaborative development of work and education*. Rotterdam: Sense Publishers.

Foreword

In the Current Economy, Preserving Occupational Safety and Health Calls for Expansive Learning

The rapid pace of change and the complexity of current production systems create occupational safety and health problems that are difficult to foresee, define, and prevent (Dias and Lima 2014). They challenge the established forms of safety work and call for an expansive reframing of its object as an aspect of the overall collaborative mastery of production. Such a reframing highlights the importance of practitioners' individual and collective work-related learning.

The collaborative mastery of an activity is constantly challenged by changes in the activity itself, as well as in the conditions in which it is carried out. Currently, however, the very principles of organizing and mastering production are also undergoing an intense transformation. During the five decades of the triumph of mass production after the Second World War, a certain view of the correct way to organize production and work became self-evident common sense. Although the deployment of new digital information and communication technologies started already in the early 1970s, when the chip was invented, these new technologies have begun only recently to profoundly question the mass production methods of organizing (Perez 2010). Although the forms of organizing and mastering production based on the new technologies are now rapidly evolving, the legacy of the mass production period thinking still strongly affects people's minds.

In the evolution of mass production, tradition-based, craft-like forms of mastering production were replaced by a new form based on the functional division of labor and standardization. Functional specialists used measurements and comparisons to determine "scientifically" and on the basis of evidence the best ways to reach specific objectives. Since the late 1970s, a number of revisions of the original mass-production paradigm have evolved, which have highlighted worker initiative and participation, especially in quality deployment. Prominent among these has been the Toyota's system of continuous improvement, in which small groups of workers first identified problems in current standards, then created and tested modifications that would subsequently be strictly observed in production

(Ohno 1988). Standards and the inspection of their observance at work units have also been central instruments of safety work (Vilela et al. 2018). Also, participative development of standards has been applied in it (Pehkonen et al. 2009). The form of learning based on the improvement of standards produces empirical, stabilizing knowledge that supports the mastery and incremental improvement of the current form of production, but not its qualitative transformation.

In the implementation of new digital technologies, tasks and processes were first automated within traditional organizational structures. It turned out, however, that the new opportunities that the digital technologies offered could not be effectively exploited within single organizations, functions, and units. The traditional firm-centric way of organizing production became increasingly superseded by new process-, network-, and platform-based forms, in which work, and information flowed across the boundaries between enterprises' units and between enterprises themselves within complex production networks (Bodrožić and Adler 2018). Here, change is continuous, because the information technology (IT) revolution is still continuing and because competition is increasingly international and innovation oriented. The development of these new kinds of production systems is a contradictory process, in which not only old and new structures, practices, and views collide, but also competing requirements and pressures have to be dealt with (Dias and Lima 2014).

Neither production or the occupational safety and health of the people working in these complex production systems can be adequately mastered with the function-centered form of learning of the mass-production period. Although functional specialization and standards continue to be important, the main challenges of learning and mastering production do not lie in reaching specific objectives within separate functions: They increasingly lie in mastering the interplay between several objectives and resolving the contradictions between them. In a functional organization, however, problem-solving usually begins by classifying problems according to their external appearance in the organizational functions and labeling them "economic," "production," "safety," etc., issues. The causes of the problems and possible solutions are then sought from the perspective of the function, which limits the possibilities for understanding the systemic causes of the problems and finding expansive solutions to overcome them.

A productive activity always has to meet many requirements, some of which appear in the current form of the production activity as contradictory. For instance, in early mass production, low unit costs and high quality were regarded as conflicting objectives, impossible to reach together. Managers thus made compromises between them – often prioritizing low costs and disregarding quality. However, when Toyota invented the system of continuous improvement, it became possible to lower unit costs and improve quality at the same time. Thus, the real challenge in learning and development in production is finding a way with which to resolve the contradiction between two crucial requirements.

The systemic causes behind frequent disturbances and accidents in production can always be traced back to aggravated contradictions within the activity system. The contradictions also cause excessive pressure and stress for those carrying out the activity. Such root causes of problems in an activity system cannot be eliminated through incremental improvements to the way in which the current principle of the activity is realized, because they lie in the very principle itself. A new principle has to be constructed, one which resolves the contradiction. For this, however, a form of learning is needed that fundamentally differs from the function-centered one of the mass-production paradigm; a type of learning that produces knowledge regarding the systemic causes of experienced problems and the possibilities for developing the system further to overcome them.

Expansive learning is precisely such a form of work-related learning (Engeström 2015). It is learning that takes place within activity systems, as the joint creation of new objects and patterns of activity, to overcome the contradictions in the present one. It is carried out as a collaborative learning activity, in which those involved in accomplishing the productive activity in different positions take jointly actions to find facts about the system of their productive activity, understand its current developmental phase and change it. Such expansive learning actions comprise questioning the current, inadequate practices and ways of thinking; analyzing the historical and systemic roots of problems in internal contradictions within the activity system; constructing a new model and principle for carrying out the activity; and implementing and stabilizing the new form. This process of joint expansive learning is characterized by a multi-voiced struggle, negotiation, and hybridization between alternative perspectives.

Although expansive learning can, in the pressure of contradictions, take place spontaneously, it typically calls for external support. Researchers can initiate and support it through a formative intervention in which they provide practitioners with the necessary conceptual instruments for taking jointly expansive learning actions and guide them in their use. This kind of developmental research carried out in collaboration with the practitioners produces knowledge about the inner contradictions behind visible problems and about realistic possibilities for overcoming them and thus improving the mastery of the activity and the occupational safety and health of those involved in it.

The Change Laboratory is a method for such collaboration between researchers and practitioners in workplaces. It differs from the intervention methods that evolved during the mass-production period in its open character. Learning in it does not begin with a predefined objective or problem and it is not limited by the boundaries of a specific function. Instead, it begins from an analysis of the historical roots and systemic context of the problems and conflicts of motives that the practitioners experience in their daily work. Its objective is to reveal the underlying systemic contradictions that cause these problems. The aggravated contradictions within the system that the analysis has revealed define the knowledge that the practitioners lack and need to find and construct: A way to resolve the contradictions. The name "expansive learning" comes from the fact that such knowledge is best found by broadening the view of the object and purpose of the activity. The result of this research and learning process is not a broadly applicable standard solution, but a new generative principle and practical method for overcoming a problem-generating contradiction within a form of production.

The method of continuous improvement created at Toyota provides a historical example of uniting opposites (Ohno 1988). This practical method enabled the organization to simultaneously lower unit costs, improve quality, and increase flexibility in the production of cars. It became the germ cell of a new form of flexible mass production, as its implementation led to further innovations that eventually completely transformed the way in which cars were produced. However, while resolving the contradiction between unit costs and quality, the new system also intensified work aggravating the contradiction between efficiency and workers' occupational safety and health.

In an expansive learning process, the collaborative analysis of the root causes of the problems that practitioners experience in their daily work, the joint construction of a way in which to resolve the contradiction behind problems, and through this the creation of a convincing, motivating perspective for development together create a shared agency for practitioners to transform their activity system.

For people accustomed to the functional form of problem-solving and learning, the Change Laboratory method's open exploratory approach may initially feel strange. For instance, in the context of the prevention of occupational accidents and diseases, practitioners often expect researchers to focus in the collaboration with them on finding immediate remedies for the occupational safety and health problems they have identified instead of analyzing their roots in the activity system. However, in the course of the expansive learning process, they see the utility of the broader, systemic approach and its capacity to create a deeper understanding, and creative solutions that initiate and support further learning and development.

February 2019

Center for Research on Activity,
Development and Learning – CRADLE,
University of Helsinki,
Helsinki, Finland

Jaakko Virkkunen

References

Bodrožić, Z., & Adler, P. S. (2018). The evolution of management models: A Neo-Schumpeterian theory. *Administrative Science Quarterly, 63*(1), 85–129. https://doi.org/10.1177/0001839217704811.

Dias, A. V. C., & Lima, F. P. A. (2014). Work organization and occupational health in current capitalism. In M. G. Ribeiro (Ed.), *Frontiers in occupational health and safety* (pp. 3–31). Sharjah: Bentham Science Publishers. https://doi.org/10.2174/97816080594851140101.

Engeström, Y. (2015). *Learning by expanding. An activity-theoretical approach to developmental research*. New York: Cambridge University Press.

Ohno, T. (1988). Beyond large-scale production. In *Toyota production system*. Portland Oregon: Productivity Press.

Pehkonen, I., Takala, E. P., Ketola, R., Viikari-Juntura, E., Leino-Arjas, P., Hopsu, L., et al. (2009). Evaluation of a participatory ergonomic intervention process in kitchen work. *Applied Ergonomics, 40*(1), 115–123. https://doi.org/10.1016/j.apergo.2008.01.006.
Perez, C. (2010). Technological revolutions and techno-economic paradigms. *Cambridge Journal of Economics, 34*(1), 185–202. https://doi.org/10.1093/cje/bep051.
Vilela, R. A. G., Jackson Filho, J. M., Querol, M. A. P., Gemma, S. F. B., Takahashi, M. A. C., Gomes, M. H. P., et al. (2018). Expansion of the object of surveillance for occupational accidents: history and challenges underwent by a reference center aiming at prevention. *Ciência & Saúde Coletiva, 23*(9), 3055–3066. https://doi.org/10.1590/1413-81232018239.21952016.

Preface

The aim of the present book is to describe Brazilian experiences of application of the Change Laboratory (CL) method within the context a research group project pursuing the development of occupational health and safety (OHS). The CL is a formative intervention method, initially developed in Finland in the mid-1990s by researchers from Center for Research on Activity, Development and Learning (CRADLE). From that time onward, this method was applied in several countries to activities in different sectors. This method was adopted quite recently in Brazil. To the best of our knowledge, the first experience dates from 2012, when it was introduced to our research group.

This innovative method was explicitly designed to promote expansive learning and development of work. These goals are achieved by encouraging participants in CL to identify and analyze historical and systemic contradictions underlying disturbances, conflict, and anomalies (such as problems relative to the environment, OHS, quality and productivity, and labor relations). The CL method is based on the activity theory approach, which includes theories, models, and concepts, such as the expansive learning cycle formulated by Engeström, which involves a series of learning actions, including questioning, analyzing, modeling, testing, implementing, and consolidating new activity system models.

Formative and interventionist research poses a challenge to contemporary science, as much as the dominant approaches to learning and development focus on diagnosis, with a clear division of labor in which solutions are developed by experts and then transferred to users. Such approaches are not always able to trigger internal mobilization, generate learning, the necessary leadership, or involvement of local actors, who are at the forefront of the work processes, thus limiting the implementation of the suggested solutions.

The present book describes several interventions inspired in the CL method to promote organizational learning in worksites with high work accident rates or other OHS disturbances.

Part I describes the trajectory of our research group, the theoretical framework for the CL method, and the effects of the financialization of productive systems on workers' health.

Part II comprises ten chapters, corresponding to ten interventions based on the CL method. Such interventions took place in several contexts, such as railway transport, social-educational assistance for adolescents, public elementary education, the hemodialysis unit of a public hospital, child labor, a school health center, a healthcare network for injured workers, urban cleaning services, and occupational health surveillance. The cases are presented according to the phase of expansive development in which the target activity was at the time of the intervention.

Part III discusses the contributions of the CL method to the techniques for accident analysis used by our research group (also known as Model of Analysis and Prevention of Accidents – MAPA). The last chapter compares the cases described and analyzes the achievements of and challenges to the application of CL in Brazil.

The authors have different academic and professional backgrounds, such engineering, medicine, sociology, psychology, nursing, psychology, occupational therapy, and physical therapy. This group works articulately and collaboratively, involving different higher education, research, and occupational health institutions.

The first two chapters' authors are professors leading the Change Laboratory within the context of Brazilian OHS. Chapter 3 was written by some well-known experts in OHS interventionist research. The authors of Chaps. 4–13 participated as researcher-interventionist in the described case studies. Chapter 14 was written by users of the MAPA method. Finally, the authors of Chap. 15 are researchers who directly or indirectly supervised or oriented the interventions.

We encouraged the authors to include brief descriptions of the context of interventions, the object of the activity which was the target of intervention, the type of data which were collected, hypotheses on historical contradictions, the double stimulation used, the learning actions developed by the participants, and the main difficulties met upon applying the method. We also encouraged them to describe innovations, variations, and possible learning outcomes.

This book targets not only researchers and students interested in OHS, formative intervention, and activity theory, but also all those interested in the development of work. We believe that it will be useful to specialists in OHS at organizations, to Ministry of Labor inspectors and Labor Prosecutors, as well as to other OHS professionals.

The ultimate goals this book are two. First, to raise awareness on the CL method in OHS practice, which includes applied research, policy development, and decision-making. By emphasizing practical application, rather than just bringing a theoretical academic discussion, we hope that this book will serve as a useful tool to support debates and dialogue on the challenges met in developing OHS.

Then, we expect that this book will contribute to the development of the participatory research, and especially of the CL method as a formative intervention. The experiences described show that at the beginning of CL sessions, the expectations to solve problems in depth are extremely high. The Brazilian context is different from the Finnish, where this method was developed, and we faced some difficulties during the process of applying CL. By considering such difficulties, we hope that this publication will help in the analysis of essential aspects of negotiation and planning, possibilities of innovations, the actual process to develop learning actions, precautions to be taken, and, of course, some aspects that we consider essential for successful interventions.

São Paulo, SP, Brazil

Rodolfo A. G. Vilela
Marco A. P. Querol
Sandra L. Beltran Hurtado
Gislaine C. O. Cerveny
Manoela G. R. Lopes

Acknowledgments

We would like to thank all the workers who made these researches possible, and the staff at Workers Health Reference Center (CEREST) in Piracicaba, which supported our fieldwork.

This publication is a result of the thematic project: "Work Accidents: From Socio-Technical Analysis to the Social Constructions of Changes." We thank all the professors, graduate fellows, volunteer researchers, and technicians in the group, specially to Universidade Estadual Paulista (UNESP FMB Botucatu), University of Campinas (FCA UNICAMP), Federal University of Minas Gerais (UFMG), Federal University of Itajubá, and Federal University of Piauí (UFPI), without whose efforts this book would had never been written.

We are also grateful to the Labor Prosecutor Office – 2nd and 15th region, São Paulo – especially to the prosecutors Dr. Mario Antônio Gomes, Dr. Silvio Beltramelli Neto, Dr. Tatiana Bivar, and Dr. Tatiana Campelo – for their support in getting access to some of the organizations which participated in the interventions.

The studies published in this book were funded by the São Paulo Research Foundation (FAPESP – Thematic project), the Brazilian Agency of Support and Evaluation of Graduate Education (CAPES), the National Council of Scientific and Technological Development (CNPq), and the Santander Group.

Contents

About the Authors

Alessandro José Nunes da Silva received his bachelor's degree in Physical Education from the Federal University of São Carlos (UFSCar). He holds a specialization in Ergonomics from the *Faculdade de Tecnologia de Piracicaba* (FATEP), a specialization in Quality of Life and Physical Activity from the University of Campinas (UNICAMP), and a master's degree in Collective Health from the Medical School of Botucatu (FMB), São Paulo State University (UNESP). He currently works at the Workers' Health Reference Center (CEREST) Piracicaba, SP, Brazil. He has experience in collective health, with emphasis on workers' health, occupational accidents, and ergonomics.

Amanda Aparecida Silva-Macaia holds a bachelor's degree in Occupational Therapy from the Federal University of São Carlos (UFSCar). She received her master's degree in Public Health and her DSc from the School of Public Health (FSP), University of São Paulo (USP), both with emphasis on workers' health. She is completing her postdoctoral research at the same institution in the following topics: organization of production processes and workers' health, analysis, and prevention of risks related to work. She works as a researcher and teacher in the field of collective health, workers' health, and Change Laboratory.

Ana Valéria Carneiro Dias received a bachelor's degree in Production Engineering, a master's degree and a PhD in Production Engineering from the Engineering School (POLI), University of São Paulo (USP). She is an associate professor at the Department of Production Engineering, Federal University of Minas Gerais (UFMG), Brazil, and invited researcher at the University of Bordeaux (2019). She is researching work organization, new forms of work organization, innovation and work organization, and the relationship between production financialization, innovation, and work.

Ana Yara Paulino obtained her bachelor's degree in Social Sciences from the Pontifical Catholic University of São Paulo (PUC-SP) and her master's degree in Political Sciences from the University of São Paulo (USP). She was researcher and educator at the Interunion Department of Statistics and Socioeconomic Studies (DIEESE) for over 20 years (1996–2017). Currently, she is a DSc candidate at the School of Public Health (FSP), USP, and professor at Strong-Esags/Getúlio Vargas Foundation (FGV), Brazil. Her topics of interest are workers' health, decent work, slave labor, and human rights.

Bianca Gafanhão Bobadilha received her bachelor's degree in Psychology from the Federal University of São Paulo (UNIFESP). She is currently a master's student at the School of Public Health (FSP), University of São Paulo (USP). Her main lines of research are mental health related to work, sexual division of labor, and Change Laboratory.

Cristiane Parisoto Masiero graduated in Odontology at the *Faculdade de Odontologia de Bauru* (FOB), University of São Paulo (USP). She holds a specialization in Workers' Health and Human Ecology from the Oswaldo Cruz Foundation (FIOCRUZ), a specialization in Permanent Education in Health in Motion from the Federal University of Rio Grande do Sul (UFRS), and a master's degree in Public Health from the Medical School of Botucatu (FMB), São Paulo State University (UNESP). She worked at the Workers' Health Reference Center (CEREST), Bauru, for 15 years, and she was a medicine preceptor at USP, and nursing and physiotherapy preceptor at the *Universidade do Sagrado Coração* (USC) at the National Council for Research and Technological Development (CNPq) in the City Hall of Bauru. She is currently a PhD student at the *Université du Quebec à Montreal* (UQÀM) in Santé et Societé. Her main lines of research are mental health related to work, occupational accidents, Change Laboratory, and violence at work.

Eugenio Paceli Haten Diniz received his bachelor's degree in Industrial Electrical Engineering from the Federal Center for Technological Education of Minas Gerais (CEFET-MG). He holds a specialization in Safety Engineering, a master's degree in Production Engineering with emphasis on Ergonomics and Health, and a PhD in Public Health with emphasis on Epidemiology from the Federal University of Minas Gerais (UFMG). Currently, he is a researcher at the Occupational Safety and Medicine Foundation, Minas Gerais (FUNDACENTRO-MG), Brazil; member of the Center for Studies and Practices in Workers' Health and Safety (CEPRASST), UFMG, and the connecting project: knowledge about work; professor and supervisor in the specialization course of Ergonomics at the Department of Production Engineering (DEP), UFMG. He develops studies and research on social innovations and interventions in workers' health and safety, analysis of occupational accidents and incidents, Return of Experience, and Ergonomic Work Analysis (EWA).

Flora Maria Gomide Vezzá graduated in Physiotherapy from the Federal University of São Carlos (UFSCar). She holds a specialization in Ergonomics from the University of São Paulo (USP) and the *Conservatoire National des Arts et Métiers* (CNAM); a master's degree in Production Engineering from the Engineering School (POLI), USP; and a DSc from the School of Public Health (FSP), USP. She is currently responsible for the ergonomics services of the São Paulo Social Service Industry (SESI-SP), Brazil. Her area of expertise is physiotherapy in workers' health, with emphasis on ergonomics, working mainly on the following topics: ergonomics, rehabilitation of occupational disorders, Ergonomic Work Analysis (EWA), management of ergonomics, musicians' health, body and work, and accessibility.

Francisco de Paula Antunes Lima received his bachelor's degree in Mechanical Engineering from the Federal University of Minas Gerais (UFMG). He holds a specialization in Safety Engineering from the *Conservatoire National des Arts et Métiers* (CNAM-Paris), a master's degree in Mechanical Engineering from the Federal University of Santa Catarina (UFSC), and a PhD in Ergonomics from the CNAM. He completed his postdoctoral research in Ergology at the *Université de Provence* and in Ergonomics at the *Université Lyon 2*. Currently, he is a full professor at the Department of Production Engineering (DEP), UFMG, Brazil, and coordinator of the specialization course in Ergonomics. He develops research and advice on health and work, safety at work (safety culture and accident analysis), work transformation, ergonomics of new technologies (ergonomics of information technology), training and skills development, economies alternatives (social economy, gift economy, economy of functionality, and cooperation), and sustainability (wastepickers, recycling, and selective collection).

Frida Marina Fischer holds a bachelor's degree in Biological Sciences from the University of São Paulo (USP); a specialization in Ergonomics from the Institute of Psychology, USP; a master's degree and a PhD in Public Health from the USP. She completed her postdoctoral research at the Institute of Occupational Health (Dortmund, Germany). Currently, she is a full professor at the Department of Environmental Health, School of Public Health (FSP), USP, Brazil; editorial board member of several scientific journals; board member of the International Commission on Occupational Health (ICOH) in the triennium 2018–2021; and a productivity fellow from the National Council for Scientific and Technological Development (CNPq, 1B). Her area of expertise is occupational health/workers' health. Her main research areas and teaching topics are ergonomics, work organization, shift and night work, early functional aging, return to work, and professional rehabilitation.

Gislaine Cecília de Oliveira Cerveny received her bachelor's degree in Physical Therapy from the Methodist University of Piracicaba (UNIMEP), her master's degree in Physiology, and her DSc from the University of Campinas (UNICAMP). She was a professor at the UNIMEP from 1987 to 2013. She has experience in the area of physical therapy, with emphasis on public health, preventive physiotherapy, working currently in research with ergonomics and Change Laboratory.

Ildeberto Muniz de Almeida graduated in Medicine at the Federal University of Espírito Santo (UFES). He holds a master's degree and a PhD in Public Health from the University of São Paulo (USP). Currently, he is a professor at the Department of

Public Health, Botucatu Medical School (FMB), São Paulo State University (UNESP), Brazil. He has experience in the area of collective health, with emphasis on workers' health, working mainly on the following topics: occupational accidents, accident prevention, analysis of accidents, conceptions of accidents, and workers' health.

Jairon Leite Chaves Bezerra obtained his bachelor's degree in Physiotherapy from the State University of Piauí (UESPI) and a degree in Radiology Technology from the Federal Center for Technological Education of Piauí (today, IFPI). He has an improvement in Physiotherapeutic Technical Expertise. He holds a specialization in Pedagogical Training of Higher Education in Health from the University Center *Uninovafapi*, a specialization in Ergonomics and Physiotherapy from the *Instituto Educacional Veronesi* (IEDUV), and a master's degree in Public Health, Politics and Management in Health with emphasis on Epidemiology, with research line in workers' health. He is currently a DSc candidate at the School of Public Health (FSP), University of São Paulo (USP), in the area of workers' health. He is also a physiotherapist at the Family Health Support Center (NASF) of the City Hall of Parnaíba and a technologist in Radiology at the Teresina Emergency Hospital (HUT), Brazil. He has experience in the area of health and education, with emphasis on collective health, workers' health, occupational physiotherapy, and medical radiology.

José Marçal Jackson Filho graduated in Mining Engineering at the Engineering School (POLI), University of São Paulo (USP). He holds a master's degree and a PhD in Ergonomics from the *Conservatoire National des Arts et Metiers* (CNAM). Currently, he is a full researcher at the Occupational Safety and Medicine Foundation, Paraná (FUNDACENTRO-PR), and coordinator of the research group/research line "Analysis, evaluation, and intervention in health and work." He develops studies and interventions in the field of prevention of occupational accidents.

Laura Seppänen holds an MSc and a DSc in Agriculture and Forestry. She is currently chief scientist at the Finnish Institute of Occupational Health (FIOH) and an adjunct professor at the University of Helsinki, Finland. She studies participatory development of work and workplace learning in various domains such as social and health care, agro-food systems, rail traffic control, and correctional services. Laura's recent research includes service networks and robotic surgery, and she is particularly interested in new digitalized forms of work. She has international collaboration in Brazil, Switzerland, and Norway. Laura is an editorial board member of the journal *Outlines: Critical Practice Studies*, and she has more than 70 scientific publications.

Luciana Pena Morgado received her bachelor's degree in Sociology from the *Fundação Escola de Sociologia e Política de São Paulo* (FESPS). She holds an MSc from the School of Public Health (FSP), University of São Paulo (USP), and, currently, she is a DSc candidate at the same institution. Her research topic is workers' health in a center of socio-educational care for adolescents in conflict with the law. She was technical advisor at the Interunion Department of Statistics and Socioeconomic Studies (DIEESE) from 2009 to 2015. Her main lines of research are sociology of work, workers' health, and Change Laboratory.

Manoela Gomes Reis Lopes received her bachelor's degree in Physical Therapy from the Federal University of Paraíba (UFPA). She holds a specialization in Family Health and Public Health at the *Faculdades Integradas de Patos*, an MSc and a DSc from the School of Public Health (FSP), University of São Paulo (USP). She completed her doctoral interchange at the University of Helsinki, under the supervision of Professor Yrjö Engeström. She is currently a professor at the Department of Biological Sciences and Health, Federal University of Amapá (UNIFAP), Brazil. She is completing a postdoctoral research at the FSP, USP. Her areas of expertise are public health, health surveillance, workers' health, prevention of occupational accidents and injuries, ergonomics, Developmental Work Research, and Change Laboratory.

Mara Alice Conti Takahashi holds a bachelor's degree in Sociology from the Pontifical Catholic University of São Paulo (PUC-Campinas) and a specialization in Ergonomics from the Methodist University of Piracicaba (UNIMEP) and the Federal University of Minas Gerais (UFMG). She also obtained a master's degree and a PhD in Collective Health from the Faculty of Medical Sciences, University of Campinas (UNICAMP). She completed a postdoctoral research in Environmental Health and Workers' Health at the School of Public Health (FSP), University of São Paulo (USP). She is currently completing a postdoctoral research in Public Health at the Medical School of Botucatu (FMB), São Paulo State University (UNESP).

Marco Antonio Pereira Querol holds a bachelor's degree in Agricultural Engineering from the University of São Paulo (USP), a master's degree in Management of Agroecological Knowledge and Social Change from the University of Wageningen, and a PhD from the Institute of Behavioral Sciences, University of Helsinki. He completed his postdoctoral research in innovation and sustainability at the University of Helsinki. Currently, he is a professor at the Department of Agronomics Engineering, Federal University of Sergipe (UFS), Brazil, and supervisor of postgraduates using the methodology of Change Laboratory. He has experience in sustainable rural development, Cultural-Historical Activity Theory, and Change Laboratory.

Marcos Hister Pereira Gomes obtained his degree in Safety Technic from the National Service of Commercial Learning (SENAC) Piracicaba, his bachelor's degree in Administration at the *Faculdade Anhanguera* of Piracicaba, and his master's degree in Work, Health, and Environment from the Occupational Safety and Medicine Foundation, São Paulo (FUNDACENTRO-SP). Currently, he works at the Workers' Health Reference Center (CEREST) Piracicaba, SP, Brazil. He is also a professor of the safety technical course at the Polibrasil College and supervisor of monographs of safety engineering specialization at the *Universidade Cruzeiro do Sul*. He works on the following research lines: workers' health, occupational accidents, collective health, public health, and education.

Marina Zambon Orpinelli Coluci graduated in Physiotherapy at the Federal University of São Carlos (UFSCar). She holds a master's degree in Nursing and a PhD in Health Sciences from the Faculty of Medical Sciences, University of Campinas (UNICAMP). She completed her postdoctoral research at the School of Public Health (FSP), University of São Paulo (USP). She works on the research areas of workers' health and ergonomics.

Rafael Junqueira Buralli graduated in Physiotherapy at the *Universidade de Marília*. He holds a specialization in Collective Health from the *Faculdade de Ciências Biomêdicas* (FACIMED) and a specialization in Teaching in Higher Education from the *Centro Universitário Ingá* (UNINGA). He also obtained an MSc from the School of Public Health (FSP), University of São Paulo (USP), and he is currently a DSc candidate at the same institution. He was a visiting researcher at the Department of Epidemiology, School of Public Health, University of Chile, and at the Center for Environment Research and Children's Health (CERCH), School of Public Health, University of California, Berkeley. He is especially interested in environmental and occupational health, epidemiology, exposure to contaminants and effects on human health, Sustainable Development Goals, and social participation.

Raoni Rocha Simões holds a bachelor's degree in Physiotherapy from the Pontifical Catholic University of Minas Gerais (PUC-Minas) and a specialization in Ergonomics from the Federal University of Minas Gerais (UFMG). He also obtained a master's degree in Work and Society Sciences, with emphasis on Ergonomics from the *Conservatoire National des Arts et Métiers* (CNAM-Paris), and a PhD in Ergonomics and Work Organization with emphasis on systemic safety management in high-risk companies from the University of Bordeaux, where he received a mention *Très Honorable*. Currently, he is an adjunct professor and coordinator of the Health and Safety Engineering course at the Federal University of Itajubá (UNIFEI), Itabira campus, MG, Brazil, and he is student of Production Engineering at the University of North of Paraná. His areas of expertise are ergonomics, workers' health, safety, and quality management in complex systems, human reliability at work, organizational analysis of accidents, and resilience engineering. In these subjects, he has participated in national and international research projects with companies, mainly in the oil extraction, electric power distribution, and meat processing sectors.

Renata Wey Berti Mendes graduated in Psychology from the Methodist University of Piracicaba (UNIMEP). She holds a specialization in Ergonomics from the UNIMEP and the Federal University of Minas Gerais (UFMG), a master's degree in Collective Health from the Medical School of Botucatu (FMB), São Paulo State

University (UNESP). She also obtained a DSc in Production Engineering with emphasis on Innovation Management from the Federal University of Rio de Janeiro (UFRJ), and a PhD in Sociology and Anthropology from the *Université Lumière Lyon 2*. She has 16 years of experience in workers' health at public institutions. Her main lines of research are mental health related to work, ergonomics in projects, innovation management, occupational accidents, and Change Laboratory.

Sandra Donatelli graduated in Pedagogy and Law from the Campos Salles Integrated Colleges (FICS/SP). She holds a specialization in Ergonomics from the Engineering School (POLI), University of São Paulo (USP), and a specialization in Health Education from the *Faculdades São Camilo* (FSC/SP). She also obtained an MSc and a DSc from the School of Public Health (FSP), USP. She works at the Occupational Safety and Medicine Foundation, São Paulo (FUNDACENTRO-SP), Brazil, in the area of labor analysis with a focus on the development of methodologies: Collective Analysis of Work (ACT), Ergonomic Work Analysis (EWA), Analysis and Prevention of Accidents Model (MAPA), and Change Laboratory based on Cultural-Historical Activity Theory.

Sandra Francisca Bezerra Gemma obtained her bachelor's degree in Nursing from the University of Campinas (UNICAMP). She holds a specialization in Ergonomics from the Production Engineering Department, Engineering School (POLI), University of São Paulo (USP); a master's degree and a PhD in Agricultural Engineering in the theme of Ergonomics from the UNICAMP. She has experience in the area of ergonomics, mainly in ergonomics of activity, Ergonomic Work Analysis (EWA), health and work, and complexity. She is currently a professor at the Faculty of Applied Sciences (FCA), UNICAMP-Limeira, SP, Brazil, in the area of engineering in Ergonomics, Health, and Work, and a full professor of the ICHSA – Interdisciplinary Master's Degree in Applied Human and Social Sciences of FCA-UNICAMP in the research line of sustainability and social protection.

Sandra Lorena Beltran Hurtado received her bachelor's degree in Physical Therapy from the National University of Colombia (UNAL). She holds an MSc from the School of Public Health (FSP), University of São Paulo (USP), and she is currently a DSc candidate at the same institution. Her area of expertise is occupational health, mainly on the following topics: occupational accidents, accident investigation methodologies, ergonomics, safety culture, oil and gas industry, and Change Laboratory.

Sandra Regina Cavalcante graduated in Law at the *Universidade São Francisco* and in Computer Sciences at the Pontifical Catholic University of São Paulo (PUC-SP). She holds a specialization in Environmental Law from the University of São Paulo (USP) and a specialization in Labor Law from the School of Law, Lawyers Association, São Paulo (ESA-OAB/SP). She obtained an MSc and a DSc from the School of Public Health (FSP), University of São Paulo (USP). She is lawyer in the labor and civil areas since 2002. She works currently at the Research and Extension Center “Labor beyond the labor law: dimensions of labor-law clandestinity” developed at the Law School, USP. She is a substitute member of the Research Ethics Committee of the School of Nursing (EE), USP, professor of law at the Campos Salles Integrated Colleges (FICS), and professor of legislation and supervisor of monographs in the safety engineering specialization at the *Universidade Cruzeiro do Sul.*

Rodolfo Andrade de Gouveia Vilela received his bachelor’s degree in Mechanical Engineering from the Engineering School (POLI), University of São Paulo (USP). He holds a master’s degree and a PhD in Collective Health from the University of Campinas (UNICAMP). He completed his postdoctoral research at the University of Helsinki, under the supervision of Professor Yrjö Engeström. Currently, he is a senior professor, researcher, and supervisor of postgraduates using the methodology of Change Laboratory at the School of Public Health (FSP), USP, Brazil. He has experience in public health, working on the following subjects: workers’ health policies, work-related accident prevention, ergonomics, and Change Laboratory.

Sayuri Tanaka Maeda graduated in Nursing at the School of Nursing (EE), University of São Paulo (USP) and in Economics at the Pontifical Catholic University of São Paulo (PUC-SP). She holds a master’s degree in Nursing Management and a PhD in Nursing from the USP. She completed her postdoctoral studies at the Finnish Institute of Occupational Health (FIOH), Helsinki, Finland. She is currently a full professor and researcher at the Department of Public Health Nursing, EE, USP, Brazil, and member of the research groups: “Analysis and prevention of work-related risks,” registered at the National Council for Scientific and Technological Development (CNPq) and “Technical assistance models and health promotion.”

Silvana Zuccolotto graduated in Civil Engineering at the Federal University of Espirito Santo (UFES). She holds a master’s degree in Industrial Engineering from the Pontifical Catholic University of Rio de Janeiro (PUC-RJ). She participated in the Latin America for Academic Formation Program (ALFA) at the Laboratory of Ergonomics of Work Neurosciences, *Conservatoire des Arts et Métiers* (CNAM-Paris). She also obtained a PhD in Production Engineering from the University of São Paulo (USP). She is completing her postdoctoral research at the Medical School of Botucatu (FMB), São Paulo State University (UNESP). Her area of expertise is ergonomics in transport companies and industries, transport engineering emphasizing the user service, incorporation, and improvement of the abilities developed by employees in work process.

Silvio Beltramelli Neto graduated in Law at the Pontifical Catholic University of Campinas (PUC-Campinas). He holds a specialization in Labor Law and Process from the PUC-Campinas, a master’s degree of Law from the Methodist University of Piracicaba (UNIMEP), and a PhD in Law from the University of São Paulo (USP). He is completing his postdoctoral research in Economic Development at the

Institute of Economics at the University of Campinas (UNICAMP), integrating the Center for Union Studies and Labor Economics (CESIT). Currently, he is a professor at the Center for Applied Human and Social Sciences, Law School, PUC-Campinas, linked to the Graduate Program in Law, research line "International Cooperation and Human Rights," prosecutor at the Brazilian Labor Prosecution Office (MPT), and editor-in-chief of the journal *Revista Jurídica Trabalho e Desenvolvimento Humano* from MPT – 15th region.

Susana Vicentina Costa graduated in Psychology at the *Universidade de São Marcos.* She holds a specialization in Clinical Hospital Psychology from the Heart Institute (InCor), Medical School (FM), University of São Paulo (USP), and an MSc from the School of Public Health (FSP), USP. She is currently a psychologist at the Center for Attention to Women's Health in the city of Franco da Rocha and the Basic Health Unit in the city of Mairiporã, SP, Brazil.

William da Silva Alves received his degree of Safety Technician from National Service of Commercial Learning (SENAC), Piracicaba, and his bachelor's degree in Production Engineering from the *Faculdade Anhanguera* of Piracicaba. He holds a specialization in Safety Engineering from the *Faculdade de Tecnologia de Piracicaba* (FATEP). He is currently a student of Philosophy at the *Universidade Paulista* (UNIP), Piracicaba, and a professor in safety technical courses in Piracicaba, SP, Brazil. He works as volunteer researcher in workers' health, safety technical support and accident analysis courses at the Research Group on Occupational Accidents (PesquisAT), Forum of Occupational Accidents (ForumAT), and Association for Environmental Health and Sustainability (ASAS). His main lines of research are ergonomics in projects, occupational accidents, and Change Laboratory.

Abbreviations

AS	Activity System(s)
CHAT	Cultural Historical Activity Theory
DWR	Developmental Work Research
OHS	Occupational Health and Safety
USP	University of São Paulo
UNESP	São Paulo State University
UNICAMP	University of Campinas
ZPD	Zone of Proximal Development

From the Portuguese

CEREST	Workers' Health Reference Center (Centro de Referência em Saúde do Trabalhador)
CIPA	Internal Commission on Accident Prevention (Comissão Interna de Prevenção de Acidentes)
FAPESP	São Paulo Research Foundation (Fundação de Amparo à Pesquisa do Estado de São Paulo)
FMB	Medical School of Botucatu
FSP	School of Public Health (Faculdade de Saúde Pública)
MAPA	Model of Analysis and Prevention of Accidents (Modelo de Análise e Prevenção de Acidentes)
MPT	Labor Prosecutor Office (Ministério Público do Trabalho)
MTE	Ministry of Labor and Employment (Ministério de Trabalho e Emprego)
NR	Standard Normative (Norma Regulamentadora)
SESMT	Occupational Health and Safety Department (Serviço Especializado em Engenharia de Segurança e em Medicina do Trabalho)
SUS	Brazilian Unified Health System (Sistema Único de Saúde)
TAC	Conduct Adjustment Agreement (Termo de Ajuste de Conduta)
VISAT	Occupational Health Surveillance (Vigilância em Saúde do Trabalhador)

Part I
Introduction

Chapter 1
Workers' Health: From Diagnosis to Formative Intervention

Rodolfo Andrade de Gouveia Vilela, Marco Antonio Pereira Querol, Ildeberto Muniz de Almeida, and José Marçal Jackson Filho

Abstract This chapter includes a brief history of the creation of our research group and the need for formative interventions in workers' health. The traditional approaches to work health and safety are reductionist and top-down; interventions are performed by experts and focus on compliance with rules or reduction of elementary hazards, without involvement of and learning by actors. Aiming at overcoming these shortcomings, the workers' health concept and movement in Brazil allowed for substantial advances from mere visible aspects to the understanding of interactions within organizations. Such considerable efforts further reveal the need for theoretical and conceptual improvement and pedagogic tools to include learning within interventions. To our group, the Change Laboratory (CL) method and Cultural–Historical Activity Theory (CHAT) represent a new paradigm as systemic view and a pedagogic contribution to interventions which aim at overcoming mere diagnosis and move toward transformation.

R. A. G. Vilela (✉)
Department of Environmental Health, School of Public Health,
University of São Paulo, São Paulo, SP, Brazil
e-mail: ravilela@usp.br

M. A. P. Querol
Department of Agronomic Engineering, Federal University of Sergipe,
São Cristóvão, SE, Brazil

I. M. de Almeida
Department of Public Health, Botucatu School of Medicine, São Paulo State University
"Júlio de Mesquita Filho", Botucatu, SP, Brazil
e-mail: ialmeida@fmb.unesp.br

J. M. J. Filho
Occupational Safety and Medicine Foundation Jorge Duprat Figueiredo – Fundacentro,
Curitiba, PR, Brazil
e-mail: jose.jackson@fundacentro.gov.br

R. A. G. Vilela et al. (eds.), *Collaborative Development for the Prevention of Occupational Accidents and Diseases*, https://doi.org/10.1007/978-3-030-24420-0_1

1.1 Brief History of Our Group and Reasons to Study Interventions on Workers' Health

Our group of researchers and experts in workers' health reunited over time professionals affiliated with Unified Health System surveillance, fiscal auditors of the Ministry of Labor and Employment, attorneys from the Labor Prosecutor's Office, and public university professors and investigators at Fundacentro. This reunion enabled, among other achievements, the formulation of public interventions targeting the prevention of work accidents and other occupational health problems.

We might point out two important aspects the group holds in common: the need to ensure protection to workers and their health in the workplace and commitment to effectively transform their working conditions.

As we describe below, our focus on the object of prevention led us to continuous reflection on notions, practices, and on the group itself.

Our initiatives resulted in the following achievements:

(a) Formulation and development of the rationale underlying information systems (IS) for work accidents (Vilela et al. 2001; Cordeiro et al. 2005; Assis et al. 2018). The notion of system demands consideration of the full scope of processes, including the following among other steps: data collection, development and management of databases, selection of cases for in-depth analysis, interventions to prevent the occurrence of new cases, feedback provision to participants, and application of findings to the planning of actions and IS improvement.
(b) Understanding the accidents as complex phenomena with immediate and late origins and consequences associated with the promotion of accident analysis practices based on notions from activity ergonomics (Guerin et al. 1997) and other fields of knowledge previously used in studies of major accidents. These developments resulted in the Model of Analysis and Prevention of Accidents (MAPA) which has since been used for surveillance and prevention of events (Almeida and Vilela 2010; Almeida et al. 2014).
(c) Creation and maintenance of the Work Accident Forum,[1] namely, a facility for virtual and in-person reflection on analysis and surveillance strategies and discussion of the results of interventions, difficulties, and facilitating factors in problem-solving. The Forum proved to be a promising continuing education network for professionals, in addition to behaving as the kernel of a work health and safety protection social network (Vilela et al. 2012a).
(d) Promotion of an interdisciplinary and intersectoral approach and dialogue involving many different social and institutional actors, also including representatives of workers and the work safety and health staff of companies whenever possible, within an environment devoted to scientific dialogue, search for continuous improvement, and stimulation of organizational learning.

Our interest in improving workers' health surveillance and prevention actions led us to several theoretical and practical reflections which we summarize next.

[1] The Forum actions are available at www.forumat.net.br

1.2 Brief History of the Conceptual Evolution of Workers' Health in Brazil

From the historical perspective, workers' health emerged in Brazil as a critical response to the factor-centered view inherited from hygiene and the approach to occupational health which does not take the full scope of the work process into consideration, nor systemic interactions which might explain the occurrence of work accidents and other work-related health problems (Dias 1994; Minayo-Gomez and Thedim-Costa 1997).

The workers' health approach sought to enhance autonomy based on the development of social networks and the participation of workers in the planning and management of health actions. Thus it detached itself from the notions and practices proper to the older paradigms.

Transdisciplinary and intersectoral methods—always from an all-encompassing perspective—were used for this purpose, including the participation of workers as subjects and partners able to contribute with their knowledge to advance the understanding of the impact of work on the health–disease process and effectively help transforming society (Dias 1994; Minayo-Gomez and Thedim-Costa 1997).

This theoretical–conceptual view was included in the Ministry of Health official guidelines. As an example, here we describe the case of the Workers' Health Surveillance Activity (VISAT). The corresponding guidelines, formulated by the Ministry of Health at the end of the 1990s, might be summarized as follows (Ministério da Saúde 1998):

> Continuous and systematic actions to detect, learn about, investigate and analyze factors determining and conditioning health problems related to work processes in their social, technical and epidemiological aspects to plan and evaluate interventions aiming at their elimination or control.

To achieve these goals, it is recommended considering the universality of actions independently from the employment relationship type; the integrative approach which combines prevention, promotion, and care; pluri-institutionality which shows that the object of interest is not restricted to the field of health; social control; hierarchization and decentralization of actions at the local, regional, state, and national levels; interdisciplinarity; research–intervention; and the transforming nature of interventions (Ministério da Saúde 1998).

As can been seen, the VISAT approach requires enlarging the objects of action, the resources used, involved actors, and space of action beyond mere narrow-scoped and fragmented inspection.

Our group gradually became aware that the conceptual and institutional advances made by workers' health in Brazil still lack practical and methodological tools to support the daily practice of professionals.

How can one understand and act on determinants when the analytical tools are still tied to the older paradigm centered on isolated risk factors (Vilela et al. 2012b; Costa et al. 2013)? How might promotion and prevention be performed and society involved in broader-scoped surveillance when the predominant culture is still the traditional command–control one, centered on the figure of the all-knowing expert (Vilela and Hurtado 2017)?

Since its inception, VISAT services always sought to include workers and other actors in multi-institutional sectorial actions, as, e.g., collective surveillance actions and sectorial agreements and bargaining. However, such enrichment of the participatory processes notwithstanding, actions remained restricted to the transformation of visible aspects, the ones considered in the legislation and regulations in particular (Vilela et al. 2018).

In the attempt at overcoming these shortcomings, our group sought conceptual and methodological support in activity ergonomics (Assunção and Lima 2003). This perspective enabled us to see work more clearly as a determinant of risk, i.e., the explanation of accidents and other health problems as the result of contradictions within the work process.[2]

One of the MAPA founding concepts (Almeida and Vilela 2010; Almeida et al. 2014) is the understanding of usual work, since accidents remain an enigma without such comprehension. Our approach began to include the comprehension of behaviors in the workplace from the perspective of situated action. This is to say, accidents are explained as the result of reduced margins of action for operators and difficulties imposed by the organization of work on the possibility to perform adjustments able to deal with variability and resume the course of action.

Accidents came to be analyzed much beyond the visible aspects of working conditions and situations. Contributions such as the ones by Reason (1990) and Llory and Montmayeul (2010) emphasize the notion of organizational or psycho-organizational accidents, which analysis should explore four axes: *transverse*, i.e., relationships among sectors or managers; *vertical*, i.e., hierarchical and power relations among the various areas within a system; *historical*, i.e., a timeline, with emphasis on changes (of direction, strategic policies, technological and organizational bases, etc., in particular) with impact on the development of a system; and *communication* among the various hierarchical levels and different areas within a same hierarchical level.

In spite of considerable advances, our experience with activity ergonomics (Guérin et al. 1997) gave rise to new inquiries. Is "organization of work" as unit of analysis clear enough to account for the origin of disturbances which interfere with current practice? Is the situated action approach sufficient to explain the history, dynamics, and origins of disturbances in systems?

On the subjects' side, the surveillance intervention proved to depend on the interventionists, as ergonomics and MAPA still lack a theory of learning able to provide conditions and a co-development platform for all the steps of intervention

[2] Activity ergonomics makes a diffuse use of contradictions, which are not granted a relevant place in the explanation of the determinants of work or their relationship with health and safety. In turn, cultural–historical activity theory (CHAT) allows enlarging our analytic ability. Notions central to CHAT are activity system (AS) as unit of analysis, contradiction and historicity as explanatory sources for the origins (determinants) of disturbances, and anomalies in AS. A more thorough description of CHAT notions is provided in Chap. 2.

(Vilela et al. 2012b). Within this context and even when refusing the role of the expert, ergonomists and accidents analysts who use the original version of MAPA still play a central part as main actors in the process of transformation from the formulation of problems to pre-diagnosis, selection of variables, conclusions, and recommendations.

Our discussions led us to put the notions of diagnosis and recommendation into question. In his first visit to Brazil, Professor Jaakko Virkkunen stimulated our debate on the notion of diagnosis. According to him, we may only come to truly know an object when we try to transform it. This is to say, a complete diagnosis is not possible without intervention (Virkkunen et al. 2014).

We only acquire full knowledge of reality when we identify resistances to its transformation. Explanations before an intervention provide hypotheses, i.e., provisional explanations of the origin of disturbances which interfere with an activity. Who should have agency in the process of analysis, formulation of solutions, testing, and implementation of changes? Which should be the role of interventionists and the internal actors of organizations (Virkkunen and Newnham 2013) when one of the goals is to promote learning and agency by means of transformation?

Therefore, in spite of the recommended care with validation, self-confrontation, and feedback provision, local actors might not expand their learning to a higher level, namely, the transformative learning needed to change the activity model which first gave rise to work-related disturbances. They might remain as mere spectators, actors who perform narrow-scoped actions, and sources for consultation, while they do not see themselves as owners of the process of transformation (Vilela et al. 2018).

1.3 From Diagnosis to Intervention, the Search for Systemic and Formative Intervention Methods

Plotting a zone of proximal development (ZPD)[3] allows understanding the trends of expansion of the object of activity work accident surveillance and prevention (Fig. 1.1).

The first square describes the possibilities for analysis and intervention within the command–control paradigm. It combines poor agency of the involved actors with reductionist views of the object to be transformed (human or technical failures). The predominant view, especially in Brazil, is the one based on the so-called unsafe act, which came to be seen as the cause of accidents. This paradigm is related to victim blaming, an approach convenient for the defense of employers in lawsuits.

Within such approach to surveillance, learning develops as a result of punishment. One example is inspection based on regulatory standards. Among its advantages, it

[3]A more thorough description of ZPD is provided in Chap. 2. For a better understanding of expanded MAPA, see Chap. 14.

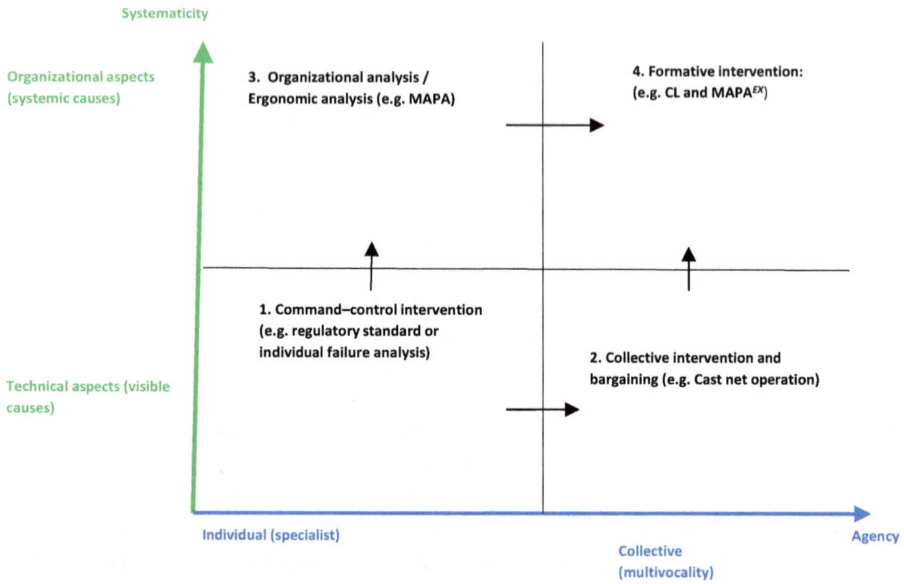

Fig. 1.1 Representation of different types of work accident and prevention intervention analysis. (Adapted from Virkkunen and Newnham (2013))

is relatively quick and easy to perform, as diagnosis consists in merely checking whether there is or not adherence to existing regulations. In addition, it allows performing interventions in a shorter time. However, besides neglecting organizational determinants, this approach exclusively focuses on deviations, but neglects their origins, and agency is restricted to experts. As a result, the possibilities of organizational learning are reduced, resulting in limitations in the ability of organizations to implement suggested recommendations, as contradictions within the management system are not discussed.

As a result, analysts end their investigation when they succeed in locating proximal causes, while the factors which led to the detected problems remain hidden. Interventions aiming at correcting the limitations of the control–command approach are collective practices such as sectorial bargaining, or the so-called cast net operation carried out by the Worker Health Reference Center (CEREST) in Piracicaba, São Paulo, Brazil[4] (second square in the figure). This intervention significantly increases the number of actors—managers and unions participate in negotiations—but its object still remains restricted to technical and visible aspects.

MAPA (third square) was developed to overcome the limitations of the first two types of interventions. It is similar to activity ergonomics interventions, which afforded a substantial expansion of the analysis of determinants (Almeida and Vilela 2010; Almeida et al. 2014). Although it is a collective strategy of analysis, a possi-

[4]MAPA was developed by researchers and professionals at CEREST, Piracicaba, São Paulo, Brazil.

ble limitation derives from the fact that agency might remain exclusive to the external experts and public health and safety professionals who implement intervention actions, while workers and company safety professionals continue mainly as sources of information or suggestions. This is to say, they are not actors with real for decision-making power.

Once they performed the full study of an organization, analysts provide feedback to actors, including diagnosis and recommendations. While this process should be validated together with workers before implementation, specialized agency tends to prevail at this point. As a result, learning tends to be restricted to the analysts. However, the hindrances to the participation of actors are not only methodological but also inherent to the cultural and political context, especially in private organizations with little room for participation, as the voice of managers prevails.

Using MAPA, the group of analysts considers a larger set of antecedent actions which might contribute to the occurrence of accidents. This change significantly expanded the potential object of interventions. However, such expansion also gave rise to the need to develop new tools to stimulate the implementation of the new activity system.

The formative intervention method (fourth square)—inspired in the Change Laboratory (CL) (Virkkunen and Newnham 2013; Engeström et al. 1996)—was developed to overcome the latest limitations. This method combines development of local agency with expansion of the comprehension of determinants, which are understood as contradictions within the activity system (AS).

For the purpose of understanding and preventing accidents, the cultural–historical activity theory (CHAT) and CL recommend adopting the following three pillars as premises: (1) an AS as unit of analysis; (2) understanding disturbances (accidents, diseases, ill feelings at the workplace, poor product and work quality, etc.) as manifestations of existing contradictions which developed over time within the targeted AS; and (3) application of expansive learning theory in a way that enables systematic inclusion of transformation and learning into the process by means of the double stimulation pedagogic strategy.

These premises represent a tremendous expansion by comparison to the older approaches, as they afford better analytical and pedagogic consistency to interventionists and all other participating actors. Such expansion allows explaining accidents no longer as the result of chance combinations of multiple factors but as disturbances arising from contradictions which develop in the course of the history of an AS. To summarize, the emphasis is on how changes occur along the history of an AS in regard to its object, subject, instruments, rules, division of labor, community, and other systems with which it interacts.

The focus of this approach is on the identification of the origin of problems, including accidents, which might express lack of adjustment among the strategies implemented by a system to face eventual changes. In other words, the world of work changes continuously, while most activities continue, thanks to the plasticity, adaptation capacity, and resilience of human beings and other system components. However, when this does not occur, but mediations among individuals (subject), instruments (e.g., resources), rules (protocols, deadlines, goals, etc.), and community are incompatible or conflicting, problems arise, which might accumulate and contribute to degradation, thus increasing the vulnerability of the system. Prevention means changing such invisible structural tension hidden behind anomalies present in everyday work.

The expanded MAPA and CL represent a special type of intervention, which consists in collective analysis dependent on the cooperation of external and internal actors with actual participation in decision-making. The fourth square in the figure retains the technical instruments introduced in the third square and combines them with the expansive learning approach. Some aspects require attention to ensure the actual progress of participants, for instance, the coordination group. More detail on the expanded MAPA is given in Chap. 14.

To unveil, debate, and collectively transform the deep origins of disturbances mean to make work more human, healthy, and sustainable. The new approach leads to a desirable, enlarged, and advanced surveillance AS.

Interventionist and systematized research poses a challenge to present-day science, as the academic and scientific tradition focuses on diagnosis, while solutions are postponed and imprisoned within university walls. Other intervention methods, such as classic consultancy—in which experts analyze and suggest solutions—are unable to trigger internal mobilization and do not yield learning nor the needed leadership or engagement of the actors at the forefront of work processes.

Mastering and adapting the CL approach to the Brazilian occupational context is the main challenge faced by the group participating in the thematic project for the past 6 years.

The main findings, learning, and challenges met by the group are described in the present book. Our overall aims are to describe experiences with the implementation of formative interventions inspired in the CL principles and assess the resulting outcomes, learning, and challenges met during actual application of this method. This book brings a summary of the Brazilian experience and detailed descriptions of all ten experiences already concluded or in progress, together with their underlying conceptual and theoretical aspects.

References

Almeida, I. M., & Vilela, R. A. G. (2010). *Modelo de Análise e Prevenção de Acidentes de Trabalho - MAPA*. Piracicaba: CEREST.

Almeida, I. M., Vilela, R. A. G., Silva, A. J. N., & Beltran, S. L. (2014). Modelo de Análise e Prevenção de Acidentes - MAPA: ferramenta para a vigilância em saúde do trabalhador. *Ciência & Saúde Coletiva, 19*(12), 4679–4688. https://doi.org/10.1590/1413-812320141912.12982014.

Assis, L. F., Fujiwara, L., Vilela, R. A. G., & Cardoso, M. R. A. (2018). Observatório Digital de Saúde e Segurança do Trabalho (MPT-OIT). In R. Mendes (Ed.), *Dicionário de Saúde e Segurança do Trabalhador. Conceitos, definição, história, cultura* (Vol. 1, pp. 825–826). Novo Hamburgo: Proteção Publicações e Eventos.

Assunção, A. A., & Lima, F. P. A. (2003). A contribuição da ergonomia para a identificação, redução e eliminação da nocividade do trabalho. In R. Mendes (Ed.), *Patologia do trabalho* (Vol. 2, pp. 1767–1789). Belo Horizonte: Atheneu.

Cordeiro, R., Vilela, R. A. G., Medeiros, M. A. T., Gonçalves, C. G. O., Bragantini, C. A., Varolla, A., et al. (2005). O sistema de vigilância de acidentes do trabalho de Piracicaba, São Paulo, Brasil. *Cadernos de Saúde Pública, 21*(5), 1574–1583. https://doi.org/10.1590/S0102-311X2005000500031.

Costa, D., Lacaz, F. A. C., Jackson-Filho, J. M., & Vilela, R. A. G. (2013). Saúde do trabalhador no SUS: desafios para uma política pública. *Revista Brasileira de Saúde Ocupacional, 38*(127), 11–21. https://doi.org/10.1590/S0303-76572013000100003.

Dias, E. C. (1994). *Atenção à saúde dos trabalhadores no setor saúde (SUS), no Brasil: realidade, fantasia ou utopia? [tese]*. Campinas: Universidade de Estadual de Campinas.

Engeström, Y., Virkkunen, J., Helle, M., Pihlaja, J., & Poikela, R. (1996). The change laboratory as a tool for transforming work. *Lifelong Learning in Europe, 1*(2), 10–17.

Guérin, F., Laville, A., Daniellou, F., Duraffourg, J., & Kerguelen, A. (1997). *Comprendre le travail pour le transformer – La pratique de l'ergonomie*. Lyon: ANACT.

Llory, M., & Montmayeul, R. (2010). *L'accident et l'organisation*. Bordeaux: Préventique.

Minayo-Gomez, C., & Thedim-Costa, S. M. F. (1997). A construção do campo de saúde do trabalhador: percurso e dilemas. *Cadernos de Saúde Pública, 13*(Supl.2), 21–32.

Ministério da Saúde (BR). Portaria n.° 3.120 de 01 de julho de 1998. *Instrução normativa sobre ações de vigilância em saúde do trabalhador no SUS*. Diario Oficial da União, 14 jul 1998. Sec 1.

Reason, J. (1990). *Humanerror*. Cambridge: Cambridge University Press.

Vilela, R. A. G., & Hurtado, S. L. B. (2017). A reading of the crisis of prevention activities: Current paradoxes and future challenges. *The Journal Cadernos Brasileiros de Terapia Ocupacional, 25*(4), 917–926. https://doi.org/10.4322/2526-8910.ctoEN01077.

Vilela, R. A. G., Ricardi, G. V. F., & Iguti, A. M. (2001). Experiência do Programa de Saúde do Trabalhador de Piracicaba: desafios da vigilância em acidentes do trabalho. *Informe Epidemiológico do SUS (Cessou em 2002 Cont. Epidemiologia e Serviços de Saúde)., 10*(1), 81–92.

Vilela, R. A. G., Almeida, I. M., Silva, A. J. N., Gomes, M. H. P., Prado, H., Buoso, E., et al. (2012a). Forum: Social network for the surveillance and prevention of work related accidents. *Work, 41*(supplement 1), 3123–3129. https://doi.org/10.3233/WOR-2012-0572-3123.

Vilela, R. A. G., Almeida, I. M., & Mendes, R. W. B. (2012b). Da vigilância para prevenção de acidentes de trabalho: contribuição da ergonomia da atividade. *Ciência & Saúde Coletiva, 17*(10), 2817–2830. https://doi.org/10.1590/S1413-81232012001000029.

Vilela, R. A. G., Jackson Filho, J. M., Querol, M. A. P., Gemma, S. F. B., Takahashi, M. A. C., Gomes, M. H. P., et al. (2018). Expansion of the object of surveillance for occupational accidents: History and challenges underwent by a reference center aiming at prevention. *Ciência & Saúde Coletiva, 23*(9), 3055–3066. https://doi.org/10.1590/1413-81232018239.21952016.

Virkkunen, J., & Newnham, D. S. (2013). *The Change Laboratory: A tool for collaborative development of work and education*. Rotterdam: Sense Publishers.

Virkkunen, J., Vilela, R. A. G., Querol, M. A. P., & Lopes, M. G. R. (2014). The Change Laboratory as a tool for collaborative transforming work activities: An interview with Jaakko Virkkunen. *Saúde E Sociedade, 23*(1), 336–344. https://doi.org/10.1590/S0104-12902014000100027.

Chapter 2
The Theoretical and Methodological Basis of the Change Laboratory

Marco Antonio Pereira Querol and Laura Seppänen

Abstract The aim of this chapter is to briefly introduce the most important principles and concepts used in the Change Laboratory (CL) method. First, we introduce Vygotsky's theory of culturally mediated human actions for understanding human agency. Second, we introduce Engeström's model of activity system as the basic unit of analysis for depicting the structure of an activity, analyzing its historical contradictions, and simulating and modeling new practices. We proceed by introducing the concept of contradictions as force of change and the theory of expansively learning for understanding the development of human activity. We continue with the principle of double stimulation and how it is applied in the CL. Finally, the variations of the method and their expected results are briefly discussed.

2.1 Introduction

In the last decades, sociological studies of intervention have taught us that interventions in human activities rarely follow a linear process. Local subjects are not passive but actors with their own strategies that respond, resist, and defend their social spaces, their cultural boundaries, and their positions of power (Long 2001). In this context, Long (2001) proposes that interventionists should understand, identify, and approach actors' local strategies so that actors themselves can appropriate, manipulate, submit, and dismember interventions.

As pointed out by Sannino (2008, 2010), resistance and subversion should not be considered as unwanted disturbances but as an ingredient and part of interventions and must have space in interventionists' methodologies. Interventions in

M. A. P. Querol (✉)
Department of Agronomic Engineering, Federal University of Sergipe,
São Cristóvão, SE, Brazil

L. Seppänen
Centre of Expertise for Development of Work and Organizations,
Finnish Institute of Occupational Health, Helsinki, Finland

R. A. G. Vilela et al. (eds.), *Collaborative Development for the Prevention of Occupational Accidents and Diseases*, https://doi.org/10.1007/978-3-030-24420-0_2

human activities are faced with actors with identities and agency, not with anonymous and mechanical responses. Therefore, intervention methodologies need to take into account people's agency if they want to be successful (Sannino 2008, 2010).

In this context, formative interventions are needed. Such interventions differ from conventional linear methodologies in at least four important aspects (Engeström 2011):

Starting point: In linear interventions the objectives of interventions are known in advance. In formative interventions, subjects are faced with a problematic situation of their life that demands analysis and the construction of a new concept, which its content the researcher does not know in advance.

Process: In conventional linear interventions, it is expected that the subject does not exert resistance. Difficulties in the execution are perceived as weaknesses that must be corrected. In formative interventions, the intervention must be negotiated, and the subjects themselves must become agents of change and embrace the process.

Outcomes: In linear interventions, the goal is to implement a predefined solution. In formative interventions, the goal is to produce hypothesis of historical contradictions within and between activities and to produce a new concept that can be used to create new local solutions.

Role of the researcher: In linear interventions, the role of the researcher is to control all variables so that there is no deviation from what was predefined. In formative interventions, the objective is to stimulate and maintain a process of expansive transformation that will be conducted and assumed by local actors.

The aim of this chapter is to provide a brief introduction of the key concepts used in a formative interventionist method called Change Laboratory (CL). This method was created in the mid-1990s (Engeström 2001; Engeström et al. 1996) by researchers from the University of Helsinki, Finland. Since its creation, the method has already been applied in a number of activities, such as in hospitals, schools, highway construction, and different countries (Virkkunen and Newnham 2013). In this chapter we propose to present the main concepts and tools used in the method in collaboration between researchers and professionals for deep or incremental superficial transformations at work.

CL can be understood as a space where new forms of work can be experienced and experimented. As a formative methodology, CL's objective is to build transformative agency so that the local actors are empowered and become agents of changes of their activities (Engeström and Sannino 2010).

This chapter starts presenting a presentation of the theory of mediation of human action and the activity system model. Next, we present the theory of expansive learning and the method of double stimulation. We continue by presenting the setting, the planning process, and the layout of the sessions. Finally, we briefly discuss the variations and the expected outcomes of the method.

2.2 Cultural Mediation of Human Actions and Agency

Vygotsky (1978) studied the cultural mediation of human actions. Cultural mediation means that the relationship between the subject and the object is mediated by cultural means or artifacts used as signs and tools. The use of these mediators allows new forms of learning based on their assimilation and their development (Vygotsky 1978; Engeström 1999). Although they do not determine the actions of individuals, mediators provide means for the construction of new courses of actions that enable them to achieve their goals in given situations.

The theory of cultural mediation is useful for understanding human agency. In activity theory, agency is understood as the ability to construct and transform one's activity. Vygotsky argues that intentional actions are mediated by artifacts. As Vygotsky stated (1997, p. 212):

> The person, using the power of things or stimuli, controls his own behaviour through them, grouping them, putting them together, and sorting them. In other words, the great uniqueness of the will consists of a man having power over his own behaviour other than the power that things have over it. Man subjects the power of things over his behaviour, makes them serve his own purposes and controls that power as he wants. He changes the environment with the external activity and in this way affects his own behaviour, subjecting it to his own authority.

Agency originates and requires the use of external artifacts. The agency relies not only on individual capacities; it relies also on access and mastery of appropriate conceptual tools and practices, as well as on social norms and on collaborative relationships existing in the community (Virkkunen 2006).

During a CL, interventionists offer a series of artifacts (models, theories, and concepts) that act as mediators and help individuals analyze the problems faced in their activity and visualize development possibilities. Mastering these artifacts empowers individuals by creating agency, that is, by building their capacity to build and transform their own activity.

2.3 An Activity as Unit of Analysis

Despite the important advancements brought by the theory of cultural mediation proposed by Vygotsky (1978), the limitation of this model is that the unit of analysis focuses on individual actions (Engeström 2001). This limitation was overcome by A.N. Leontiev (1981), which differentiates an individual action from collective activity (inserting the individual into a community). Leontiev (1981) expands the unit of analysis for a collective activity, which always consists of actions of individuals. Moreover, an activity is understood as dynamically evolving during history. The difference between an action and an activity is the result of the increasing societal division of labor and regulated through specific relations to each historical form of production (Engeström 1987, 2015).

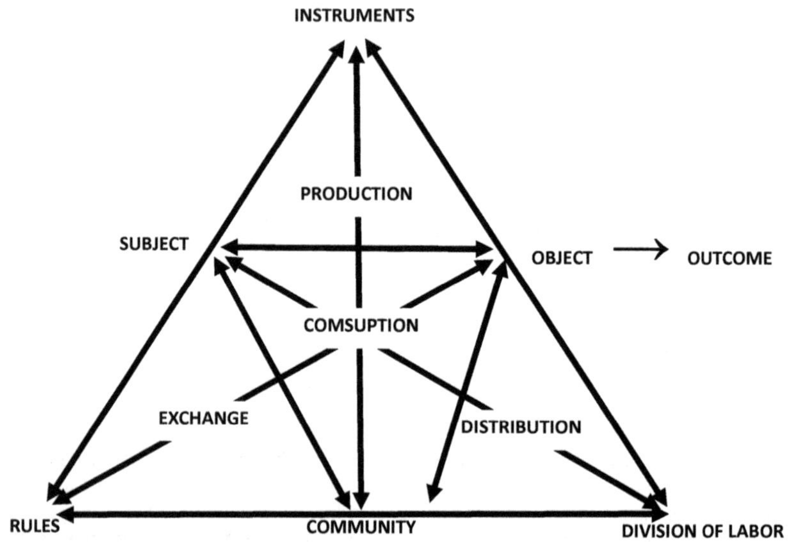

Fig. 2.1 The activity system model. (Originally published in Engeström 2015)

According to Leontiev (1978), actions are directed to goals, while an activity is always directed to an object. Therefore, the study of an activity presupposes the identification of the object that directs the actions of the subjects. Unlike the goal, which is anchored in a given place and time, the object of an activity is more sustained and open and is related to a human need (Leontiev 1978).

Engeström (1987, 2015) expanded Leontiev's ideas (1978, 1981) by discovering key mediating elements that have evolved during the history of human activities. He created a triangular graphical representation, which demonstrates the basic relationships in mediation systems of a human activity. He proposed an activity system as the basic unit of analysis, which incorporates the unit for understanding human actions. In addition to the use of cultural artifacts in the transformation of the object, the model also includes social mediators, such as rules, division of labor, and community (see Fig. 2.1).

The elements presented in this model are defined as follows:

Subject refers to an individual or group of people whose positions and views are chosen as the perspective of analysis.

Object is understood as the meaning, motive, and purpose of a system of collective activity. The object is considered by Engeström (1987, 2015) as the "raw material" for the subjects involved in an activity and may be open to numerous and partial interpretations. This raw material, which contains a certain conflict or problem to be solved, is seen as a consequence of constant interpretations, reconstructions, and modifications impelled by the action of the subjects, thus revealing the contradictory and historically changing character of the systems involved in the production of activity.

Outcome is the result of processing the object in the activity.
Tools include all instruments mediating the action of the subjects in physical or symbolic form, external and internal such as objects, furniture, text and video, electronic sound and video devices, physical space, models, plans, schemes, layout, abstractions, images, signs, and others.
Community refers to all individuals or subgroups that have the same object of the activity in question.
Division of labor is the division of tasks between the subjects.
Rules refer to regulations, norms, and conventions related to the context of the activity that are presented tacitly or explicitly. Rules may inform about their advantages, possibilities, virtues, etc., as well as the impediments, obstacles, barriers, and others that condition and restrict the actions and interactions of the subjects of the activity within the activity systems in which they are submitted.

In the CL, the activity system can be used in the simulation of activities aiming at, for example, (a) to analyze the development and formation of the elements of the activity, (b) to reveal the contradictions within and between activity systems, and (c) modeling or designing a new concept of the activity in order to solve the internal contradictions, changing the elements of the system.

2.4 Networks of Activity Systems

Activity systems are never isolated but incorporated into the network of interdependent activity systems that produce the elements and consume their products, affecting and being affected by other activities (Engeström, 1987; Pereira-Querol 2011). One way to model it is to choose an activity system as a central one and then observe all the "neighboring activities" related to that central activity system (see Fig. 2.2).

Engeström (1987, 2015) classified these neighboring activities into four groups: object activities, which are composed of all activities where the objects and results immediately emerging from the central activity are incorporated; instrument-producing activities, which include all activities that produce the main instruments for the central activity; activities producing subjects, which involve activities such as education and schooling for the central activity; and finally, rule-making activities, which include activities such as administration and legislation.

The second form of modeling the relation between activities is a network of activities described by Engeström (2001) as the model of co-production or co-configuration. In this model, there are two activity systems co-producing a partially shared object as a minimal unit of analysis (see Fig. 2.3). Activities are directed to different objects, but in this model, objects overlap, allowing collaboration to take place. Such collaboration is possible because of a shared object.

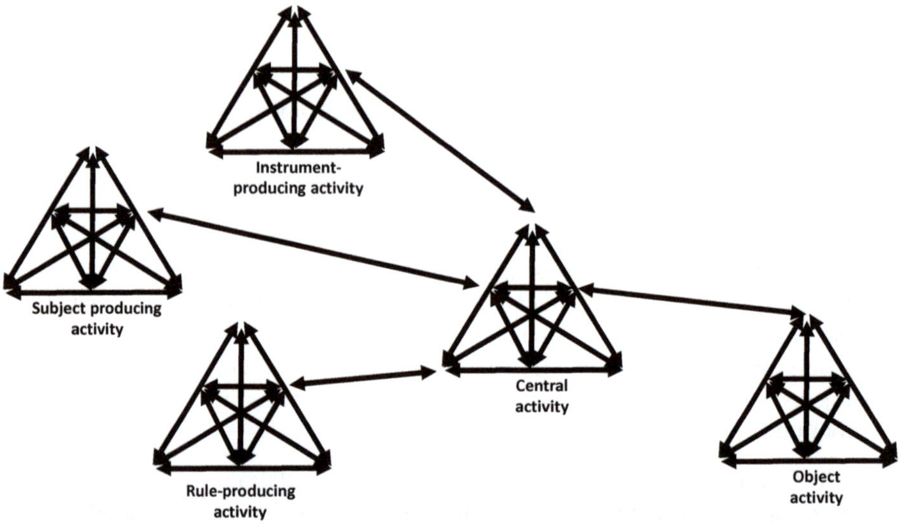

Fig. 2.2 Network of functionally interlinked activities. (Originally published in Engeström 2015)

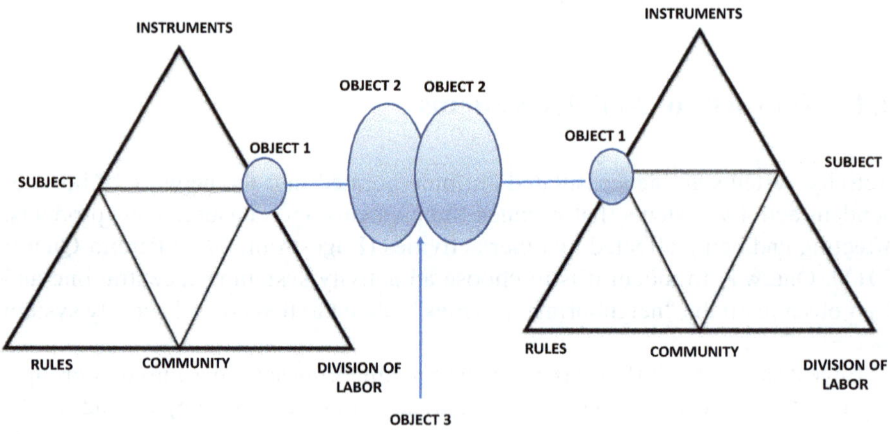

Fig. 2.3 Model of two activity systems with a partially shared object. (Originally published in Engeström 2009)

2.5 Contradictions as a Force of Development

In activity theory,[1] contradictions are the source of change and development. According to Engeström (1987, 2015), all elements of an activity system experience primary contradictions. A contradiction is not the same as a conflict or a

---

[1]The approach based on works of Vygotsky, Leontiev, and Engeström is referred to as cultural-historical activity theory, or simply activity theory.

problem (Engeström and Sannino 2011). For Engeström (2001), contradictions are "tensions within and between activity systems" (p. 137), whose origin is historical due to the double aspect of activities, which are independent and subordinated to other activities. Disturbances, conflicts, and ruptures are understood as manifestations of historical contradictions within and between the elements of activity systems.

In a capitalist system of production, the primary contradiction is expressed through a contradiction between the use value and the exchange value of a product (Marx 1990). Use value refers to utility, that is, the capacity of something to satisfy a human need. Exchange value refers to the fact that something can be delivered in exchange for something in a social context. Shelter provided by homebuilding is use value, while the price of it is the exchange value. These two values are contradictory in the sense of being mutually exclusive, although they usually coexist.

Attempts to resolve contradictions make an activity move from one stage of development to another. At each stage, the contradiction in an activity system is expressed differently. As activity systems are open, connected to many other systems, an external change (e.g., a new law or a new technology) can lead to the appearance or aggravation of a secondary contradiction. A secondary contradiction is defined as an incompatibility between two elements of an activity system. Secondary contradictions manifest in the form of disturbances and conflicts, experienced by subjects.

During a formative intervention, the participants are stimulated to formulate hypothesis of the contradictions that affect their activity. For this, a historical analysis of the system is necessary. It is important that interventionists identify the most important and aggravated contradictions because they are the ones that will motivate individuals to make changes.

2.6 Expansive Learning

The Change Laboratory is based on the theory of expansive learning, which is characterized by the expansion of the object of an activity. Such process can occur in five qualitatively different phases (see Fig. 2.4). (Engeström, 1987; Engeström 2015; see Engeström and Sannino 2010, for a recent review of studies). At each phase, a different kind of internal contradiction drives development.

In the cycle of expansive learning, the creation of new objects starts from an already consolidated activity that begins to present problems (Engeström and Sannino 2010). This phase, called the need state, is characterized by a situation of dissatisfaction with the current situation, a state of crisis, and an urgent need to change something. From this, contradictions appear in their basic form, as primary contradictions between use value and exchange value (Marx 1990). During this phase, practitioners begin to discuss and question the motive of the activity

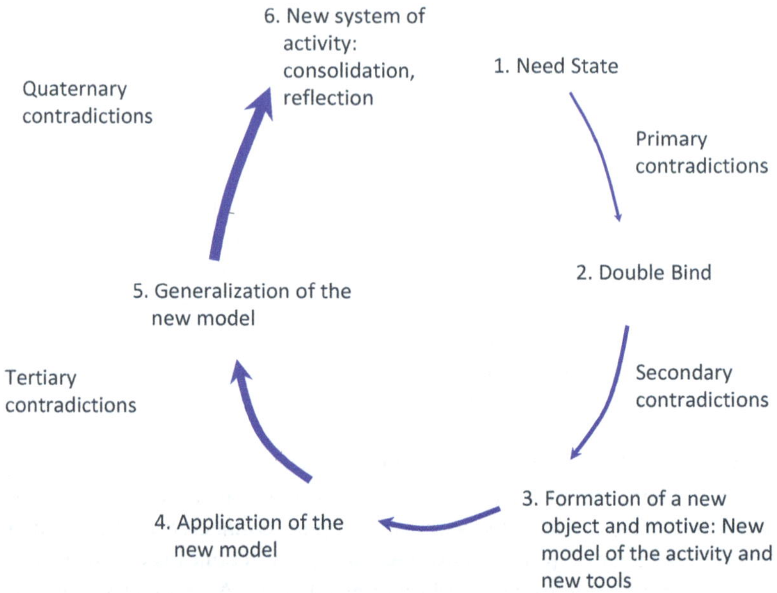

Fig. 2.4 The cycle of expansive learning. (Originally published in Engeström 2015)

(the object) and the current ways of how to obtain it (the tools), such as the methods and the technologies (Engeström 1987, 2015).

The second phase, called double bind, is characterized by the aggravation of the problems that were already being experienced in the previous phase. In this phase, new disturbances emerge, threatening the production of desired results. At this phase the contradictions evolve and develop into secondary contradictions—tensions and mismatches between the elements of the activity system. People who participate in the activity feel that it is not possible to continue doing things the way they were doing before, but they still do not know what needs to be done to solve the problems (Engeström 1987, 2015).

The aggravation of the problems leads the practitioners to seek solutions that may or may not give way to more expansive objects. An object can be considered more expansive when it has qualitatively broader characteristics than the previous one and can resolve the contradictions that affect the activity system. If the crisis is severe enough, people may question the activity system, including the reason for the activity as a whole (the object).

If the subjects challenge the object/motive of the activity and redesign it by creating a new object for their activity more expansively, then this can be called an expansive cycle. This phase is called *object construction*. In this phase, a new object and corresponding other elements of the activity system are designed to overcome the contradictions that are taking the activity to a crisis.

Once the solution is conceived, the idealized activity begins to be implanted. This phase is called *application and generalization*, consisting of the implantation of the changes in the activity system. In it, practitioners begin to put the plans into practice in order to conduct actions to transform the new object and thus produce a desired result. Misalignments between elements of the new expanded activity and elements of the prior activity begin to appear, which are termed tertiary contradictions. These mismatches may be caused by insufficient development of the new elements (Engeström 1987, 2015).

During the implementation of the new concept of the activity, it is very likely that it will begin to clash with the neighboring activities that still follow the old production logic. Therefore, before or during the consolidation, the new activity must resolve these tensions with parallel activities, which are called quaternary contradictions. If practitioners solve them, the activity will evolve to the phase of *consolidation* of the activity. Eventually, the activity system may face new contradictions and start a new, qualitatively different cycle of expansive learning.

2.7 The Double Stimulation Method

The double stimulation method is a method initially proposed by Vygotsky (1978, p. 74–75) to evaluate psychological functions such as memory, thought, and agency. The main idea of the method is that when individuals face with problematic situations that cannot be solved with the tools and knowledge they possess, individuals seek new psychological and practical tools that allow them to solve the situation (Lektorsky 2009).

The double stimulation method consists of the application of two stimuli: The first stimulus is a task to be performed or a problem to be solved. Such task must be slightly above the capacity of the individual to solve it, so the interventionist must have a sense of the level of development of the individual and his potential. The second stimulus is a neutral artifact that has the potential to be used as a tool to solve the proposed task. Once the subject incorporates, or uses, the artifact in the resolution of the task, the structure of the operation changes (Vygotsky 1978, 1997). By neutral artifact, Vygotsky refers to a cultural artifact that can be used indefinitely, that is, the interventionist does not impose, nor suggest, the way it should be used, leaving the individual free to refuse or reinvent this artifact according to their conditions, knowledge, and perception.

The double stimulation method is an example of remediation, where the individual incorporates and transforms new tools (Engeström 2007). In other words, it is a method to change the mediation of actions and thus provide conditions for the subject to be able to conduct new actions, allowing him to overcome problematic situations.

2.8 Methodological Model of the Change Laboratory

In Change Laboratory, the general model of an activity system and the expansive learning cycle are used as psychological tools that allow professionals to reflect collaboratively on their activity. The CL can be understood as a learning activity, that is, an activity whose object is at the same time the analysis and development of a productive activity.

The method is composed of six phases, which are based on the expansive learning cycle, and consists of six expansive learning actions: questioning, analyzing, designing, testing a new model, implementing the new model, and reflecting under process and consolidating the new activity (see Fig. 2.4).

Interventions consist of a limited number of participants, between 10 and 20 representatives of activities other than the interventionist and an assistant (Virkkunen and Newnham 2013). The role of the interventionist is to present the material to be discussed and to facilitate the discussions. The assistant helps during the intervention, for example, in data collection and in the use of video projection equipment.

The process begins with collecting mirror data on (1) the status of the activity, such as, historical data on important events, (2) the current practices (the way the activity is conducted), (3) the main problems faced, and (4) the main concepts and tools used in the activity.

The function of these data is twofold: to serve as mirror of the activity for supporting participants' reflection and to help the interventionist and the participants to model and analyze past, current, and future activity. The sessions begin with an analysis of data on the problems faced daily, that is, what happens on the level of actions. Next, the past phases of the activity are analyzed, and this yields an understanding of how the actual problems have emerged during history.

Once participants have analyzed the problem, the interventionist helps them to model new solutions. Again, the interventionist offers tools to support this process. Often, new suitable tools (second stimuli) are developed dialogically during the intervention process to support learning and to finally model a new concept of activity with an expanded object. Once the new concept is figured jointly, the corresponding solutions need to be experimented with. This happens again on the level of actions (see Table 2.1).

Once tested, new sessions are organized to evaluate the experiments and reflect on the new solutions and potential problems that arise during their implementation. Thus, expansive learning is enhanced by simultaneous examination of action and

Table 2.1 Movement between actions and activity during the CL

Level	Problems	Outcomes
Activity: systemic and structural	2. Formulate hypothesis of the historical contradictions	3. Design new object and/or new model of the activity system
Actions: experienced and visible	1. Identify disturbances	4. Design and implement local solutions

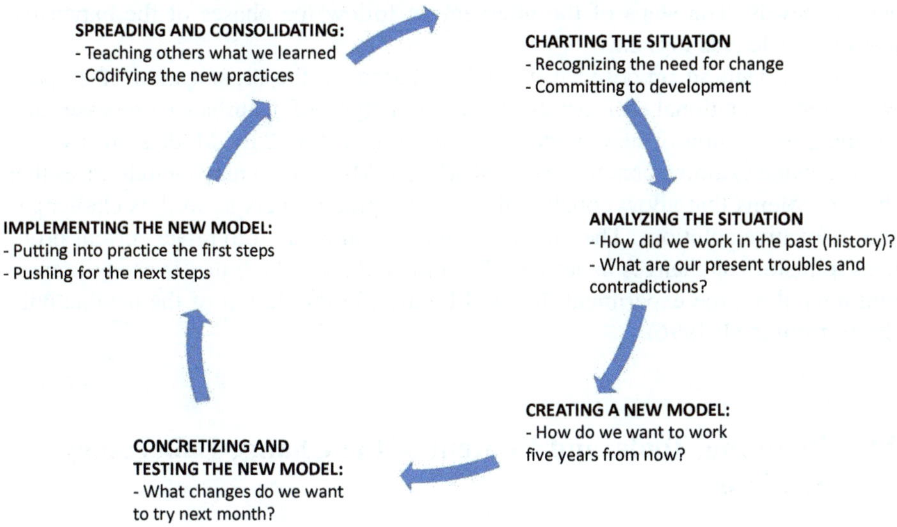

Fig. 2.5 Stages of the expansive development process during the Change Laboratory. (Originally published in Virkkunen and Newnham 2013)

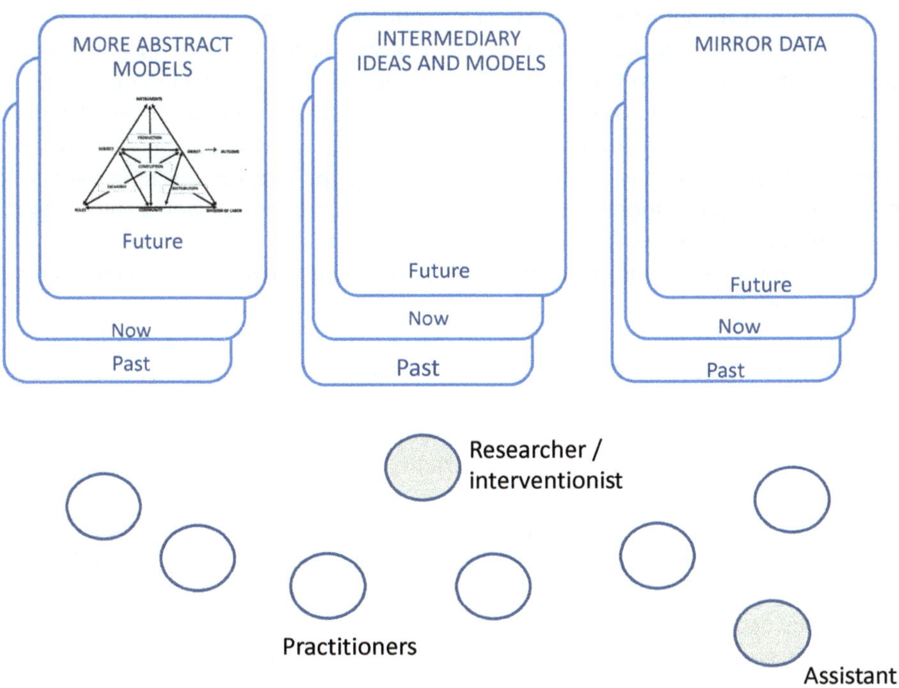

Fig. 2.6 Layout of the Change Laboratory. (Originally published in Virkkunen and Newnham 2013)

activity levels. The steps of the intervention follow the phases of the expansive learning cycle (see Fig. 2.5).

The intervention proposed in the CL is based on the organization of a space where representational tools are available for analysis of disturbance processes and for the construction of new models of activity (see Fig. 2.6). Models are used to represent and examine concrete cases of activity. Mirror data may include cases that show problems that allow collaboration among practitioners as well as challenges and innovative solutions. The image of the past can be used to relate current problems to historical changes in activity. The mirror data of the future is used to represent data about the experiment that will be used in the design of the new activity (Engeström et al. 1996).

2.9 Planning, Tools, and Content of the Change Laboratory Sessions

A CL usually begins with an analysis of the disturbances observed in the activity. Such disturbances are understood as expressions of contradictions within or between activity systems. During the CL sessions, participants produce future visions of the activity, which are not based solely on a classification of what is desired and what is not desired. In CL, the views are based an analysis of the structure of the activity system that could help to solve internal contradictions to the system in question.

Regardless of how it originates, the intervention must be formalized and detailed into an action plan. In addition to introducing the objective and strategy of the intervention, the plan also has the function of situating the intervention in the general politics of the organization (changes and political objectives), as well as inserting it in the strategy of the activity in question.

The action plan should result from the negotiation between the interventionist and representatives of the organizations involved in the intervention. One of the first steps during the planning is to define the activity to be developed during the intervention and to preliminarily outline the activity system. The second step is to understand the problem that led to the need of the intervention.

The new concept produced in the CL should not be seen as a standard solution but rather as a new concept to be materialized in a prototype to be tested and developed.

The task of modeling and implementing a new concept is best accomplished in the units that first experienced the changes and/or experienced the contradictions more strongly. The prototype should be designed and inserted preferably in a unit that has the capacity and interest to develop a new model for its activity. Generally, the existence and severity of the problem that needs to be solved by the intervention affect the participants' interest in developing solutions.

Usually there is pressure from the management of organizations to minimize the number and duration of the interventions. Therefore, it is important to discuss and negotiate the duration and timing of the intervention during the planning phase.

The selection of the participants is made by the interventionists along with a local representative of the activity to be developed (manager and/or worker). The idea is to find people whose knowledge and resources are needed for problem analysis and solution design. These people should represent different perspectives and contribute with complementary resources and knowledge. If possible, people who are highly motivated to participate and contribute to the process should be chosen. Usually the process consumes 10–12 sessions of 2–3 hours each. In addition, it is important that the sessions are sequential, i.e., without long intervals between them (maximum of 1 week).

For an intervention to be successful, the support of the organization's management is crucial, and the learning process must connect with the management of the organization. Good collaboration between managers, interventionists, and participants is necessary before, during, and after the intervention. The participants should be able to work during CL without interference from the management, but on the other hand, management and other stakeholders should be informed of their progress and intermediary results (Virkkunen and Newnham 2013).

2.10 Variations of the Change Laboratory

In each empirical setting, the CL method is adapted for the specific purpose or type of activity or network of activities. Thus, instead of being a unique method, the Change Laboratory is a set of tools that can be used and adapted to the local context. In this sense, each application of the method is a variation. However, despite the specificity of each application, some sets of variations can be identified, according to the main objective or characteristics of the activities involved in the intervention.

Among the variations of the CL, the boundary-crossing laboratory is growing importance. This variation is an application in setting in which several organizations are involved. In boundary-crossing laboratory, the unit of analysis is two or more activity systems with a shared object (Engeström 2001; Kerosuo and Engeström 2003; Kerosuo 2001, 2004). Another variation is the competence laboratory, which focuses on a learning activity of an organization (Ahonen and Virkkunen 2003). Helle et al. (2010) describe another variation applied in the setting of media. In this variation, a media concept is used as a tool and model for searching and for developing a new concept for a Finnish newspaper (called "The Paper"). Finally, there is a variation of CL called Cultural Laboratory specially designed for a setting with people from different cultures (Teräs 2007). A specific CL modification emphasizing participants' well-being, the Change Workshop, has been developed (Launis and Pihlaja 2007; Ala-Laurinaho et al. 2017).

2.11 The Expected Outcomes of a CL

The results of a CL should not be seen as a standard solution but as a new concept that is incorporated into a prototype to be tested and developed (Virkkunen and Newnham 2013). Expansive learning where new concepts of activities are created and implemented are long-term processes that may require years and several successive interventions.

Generally, an intervention is evaluated positively if the pre-established goals are met. According to Virkkunen and Newnham (2013), the task of evaluating the results of a formative intervention such as the CL is more complex because the purpose of this intervention is not only creating a change in an activity but also, and above all, the deepening of the understanding of the nature and the causes of their problems. Therefore, a successful CL intervention leads to a new and more comprehensive way of understanding the problems faced in an activity.

Reconceptualization of the causes of the problems has an important role in relation to the type of solutions produced. The deeper understanding of the causes, more radical will be the solutions, and consequently more time is needed to implement and consolidate them. Besides entire activity concepts, CL interventions may yield new or modified tools or practices for the activity.

Generally, only a small number of practical results are seen immediately after the intervention because the new concepts and solutions created during CL need time to be implemented and require some work. It is common for participants to face obstacles when trying to implement the new model. It is common to experience the contradictions between the dominant logic of activity and logic of their new concept. Unlike other types of interventions, generalizations created by CL begin locally but have the potential to become general. Therefore, diffusion occurs with evolution and enrichment, not as a direct transfer through copies of the solutions created (Virkkunen and Newnham 2013).

As mentioned above, during a CL participants use models and concepts to understand problems and design solutions. Such solutions usually require collaboration at the level of a collective activity. Therefore, CL favors the formation of transformational agency among the participants, enabling and allowing themselves to have control over the development of their activities. Another type of possible and desirable result is the formation of a learning platform, i.e., a permanent space allowing continued learning, which may start in the laboratory.

References

Ahonen, H., & Virkkunen, J. (2003). Shared challenge for learning: Dialogue between management and front-line workers in knowledge management. *International Journal of Information Technology and Management, 2*(1), 59–84. https://doi.org/10.1504/IJITM.2003.002449.

Ala-Laurinaho, A., Kurki, A.-L., & Simonsen Abildgaard, J. (2017). Supporting sensemaking to promote a systemic view of organizational changes – contributions from activity theory.

Journal of Change Management, 17(4), 367–387. https://doi.org/10.1080/14697017.2017.1309566.
Engeström, Y. (1987). *Learning by expanding. An activity-theoretical approach to developmental research*. Helsinki: Orienta-Konsultit.
Engeström, Y. (1999). Activity theory and individual and social transformation. In Y. Engeström, R. Miettinen, & R. L. Punamäki (Eds.), *Perspectives on activity theory*. Cambridge: University Press.
Engeström, Y. (2001). Expansive learning at work: Toward an activity-theoretical reconceptualization. *Journal of Education and Work, 14*(1), 133–156. https://doi.org/10.1080/13639080020028747.
Engeström, Y. (2007). Putting Vygotsky to Work: The Change Laboratory as an Application of Double Stimulation. In H. Daniels, M. Cole, & J. Wertsch (Eds.), The Cambridge Companion to Vygotsky (pp. 363–382). Cambridge: Cambridge University Press. https://doi.org/10.1017/CCOL0521831040.015.
Engeström, Y. (2009). The future of activity theory: A rough draft. In A. Sannino, H. Daniels, & K. Gutierrez (Eds.), *Learning and expanding with activity theory* (pp. 303–328). New York: Cambridge University Press.
Engeström, Y. (2011). From design experiments to formative interventions. *Theory & Psychology, 21*(5), 598–628. https://doi.org/10.1177/0959354311419252.
Engeström, Y. (2015). *Learning by expanding. An activity-theoretical approach to developmental research*. New York: Cambridge University Press.
Engeström, Y., & Sannino, A. (2010). Studies of expansive learning: Foundations, findings and future challenges. *Educational Research Review, 5*(1), 1–24. https://doi.org/10.1016/j.edurev.2009.12.002.
Engeström, Y., & Sannino, A. (2011). Discursive manifestations of contradictions in organizational change efforts: A methodological framework. *Journal of Organizational Change Management, 24*(3), 368–387. https://doi.org/10.1108/09534811111132758.
Engeström, Y., Virkkunen, J., Helle, M., Pihlaja, J., & Poikela, R. (1996). The change laboratory as a tool for transforming work. *Lifelong Learning in Europe., 1*(2), 10–17.
Helle M., Töyry, M., & Ruoranen A. (2010). *We all do our own thing*. Proceedings of the MeCCSA10 Conference:2010 January 6–8; London, UK.
Kerosuo, H. (2001). Boundary encounters' as a place for learning and development at work. *Outlines, 3*(1), 53–65.
Kerosuo, H. (2004). Examining boundaries in health care - outline of a method for studying organizational boundaries in interaction. *Outlines, 6*(1), 35–60.
Kerosuo, H., & Engeström, Y. (2003). Boundary crossing and learning in creation of new work practice. *Journal of Workplace Learning, 15*(7/8), 345–351. https://doi.org/10.1108/13665620310504837.
Launis, K., & Pihlaja, J. (2007). Changes in production concepts emphasize problems in work-related well-being. *Safety Science, 45*(5), 603–619. https://doi.org/10.1016/j.ssci.2007.01.006.
Lektorsky, V. (2009). Mediation as a means of collective activity. In A. Sannino, H. Daniels, & K. D. Gutierrez (Eds.), *Learning and expanding with activity theory*. Cambridge: Cambridge University Press.
Leontiev, A. N. (1978). *Activity. Consciousness. Personality*. Englewood Cliffs: Prentice Hall.
Leontiev, A. N. (1981). *Problems of the development of the mind*. Moscow: Progress.
Long, N. (2001). *Development sociology: Actor perspectives*. London: Routledge.
Marx, K. (1990), Capital, Vol. 1, Penguin Books, London
Pereira-Querol, M. A. (2011). *Learning challenges for sustainability: An activity theoretical study of a network from a swine industry chain* [dissertation]. Helsinki:University of Helsinki.
Sannino, A. (2008). From talk to action: Experiencing interlocution in developmental interventions. *Mind, Culture, and Activity, 15*(3), 234–257. https://doi.org/10.1080/10749030802186769.
Sannino, A. (2010). Teachers' talk of experiencing: Conflict, resistance and agency. *Teaching and Teacher Education, 26*(4), 838–844. https://doi.org/10.1016/j.tate.2009.10.021.
Teräs, M. (2007). *Intercultural learning and hybridity in the culture laboratory* [dissertation]. Helsinki: University of Helsinki.

Virkkunen, J. (2006). Dilemmas in building shared transformative agency. *Activités, 3–1*. https://doi.org/10.4000/activites.1850.

Virkkunen, J., & Newnham, D. (2013). *The Change Laboratory: A tool for collaborative development of work and education*. Rotterdam: Sense Publishers.

Vygotsky, L. S. (1978). *Mind and society: The development of higher mental processes*. Cambridge: Harvard University Press.

Vygotsky, L. S. (1997). The history of the development of higher mental functions. The collected works of L. S. Vygotsky. Vol. 4. The history of the development of higher mental functions. New York: Plenum.

Chapter 3
Work and Health and Contemporary Capitalism: Economics as a Social Disease

Francisco de Paula Antunes Lima and Ana Valéria Carneiro Dias

Abstract Work nowadays faces changes that bring important impacts on workers' health, such as accidents, large disasters, and diseases related to stress. These anomalies have been associated with raising tensions that resulted from globalization and the increase of inter-capitalist dispute. In this context emerges the phenomenon of financialization of the economy that leads organizations to want short-term results. Work management forms assume aggressive and even violent nature that stimulates competition between workers and teams by imposing ever-growing goals and loads. This context presents challenges to interventions that aim to change management models in order to make the work more humane and sustainable. The chapter addresses the following questions: What are the central contradictions in today's productive processes? How can one create a need among managers and other players for more sustained transformation of production models? What kind of demands do these changes require for intervention and surveillance activities in the context of workers' health?

3.1 Introduction

Critical studies on workers' health, whether conducted through a theoretical approach such as social epidemiology (Laurell and Noriega 1989) or those dealing with particular conflictive issues, such as outsourcing, precariousness, slave labor, child labor, and decent work, systematically reveal that economic relations determine different occupational diseases (Lima et al. 1997; Santos et al. 2009). Criticism of capitalist domination relationships appears as the main outcome of these studies, which reveal the structural contradictions that determine work and health conditions and the various forms of physical and mental exhaustion generated by the exploitation and overexploitation of labor. In this way, they update in each new productive configuration what Marx had recognized in the nineteenth century: "capitalist

F. de Paula Antunes Lima (✉) · A. V. C. Dias
Department of Production Engineering, School of Engineering, Federal University of Minas Gerais, Belo Horizonte, MG, Brazil

R. A. G. Vilela et al. (eds.), *Collaborative Development for the Prevention of Occupational Accidents and Diseases*, https://doi.org/10.1007/978-3-030-24420-0_3

production only develops the technique and combination of the social production process by simultaneously undermining the sources of all wealth: the land and the worker" (Marx 1983, p. 102). These critical approaches show, in short, how workers' diseases result from capitalist relations of production, that is, how living and falling ill are socially produced.

On the other hand, adopting a more pragmatic perspective, the disciplines that set to transform and improve working conditions (safety engineering, occupational hygiene and medicine, ergonomics, work psychology, etc.), acting based on social demands and specific problems, are faced with the limits imposed by economic systems and models, which are now being leveraged by global competition. Even technically sound interventions are ineffective against the limits imposed by production relations. We often confront the "economic blackmail" that uses the Chinese threat to maintain pathogenic working conditions in every national space. The success of the Brazilian meat industry is largely based on the sickness of thousands of workers: in some companies the dismissal rate reaches 20% of the labor force (on this subject, watch the touching documentary "Carne e Osso" – Meat and Bone (Cavechini and Barros 2011)). The serious situation has led to passing a specific act (NR-36 – Health and Safety at Slaughter and Meat Processing Work), which, however, is already unable to solve the contradictions that permeate the production system, from the competition on a global scale to the rhythm of the production line. The agricultural production records hide the sickness and death of thousands of rural workers, not to mention the consumers themselves, mainly due to the intensive use of agrochemicals (Augusto et al. 2012; Carneiro et al. 2012, 2015). Construction still have high accident rates; couriers/motorcyclists who crowd our streets continue to die in increasing numbers, most of them workers who provide services under time pressure, in poorly maintained public roads that increase the risk of accidents (Diniz et al. 2005, 2015; Silva 2012; WHO 2015). In the service sector, mental disorders compete with musculoskeletal disorders as the main cause of sick leave.

The precariousness of work in these different activities is well-known; the interventions of regulatory bodies and specialists cannot produce significant change in working conditions and improve on this globally pathogenic picture. Even in cases that are considered successful, such as civil aviation, where accidents occur on the order of one per million, these economic limits are also present, although they manifest in a subtler way. As Amalberti (2016) suggests, ultra-secure industries, such as aviation and nuclear energy, have reached limits that would require prohibitive investments to achieve significant gains in safety. These companies can differentiate between them by varying the accident rate by 1–2, but it is not possible to achieve safety jumps of order 10 from the current level, which has stabilized around 10^{-6}, unless the economic model, resulting from a worldwide organization implemented from the mid-twentieth century, changes as a whole. Referring to the relationship between safety culture and economy, he concludes:

> [...] we can significantly modify culture by arbitrarily changing the fundamentals of the technical system and introducing major changes in the system's economy, but of course this goes beyond the capabilities of a one-off intervention in an industrial or service enterprise (like a hospital or bank). In short, the market economy prescribes culture more than the

> other way around. The levers of change are systemic and not circumscribed. (Amalberti 2016, p. 134-5)

In the aviation sector, the new frontier would henceforth be to manage the post-accident instead of trying to reduce them. Punctual changes, he argues, can generate counterproductive effects, as in the case of blood control in France, when stricter requirements after a contaminated blood case led to more deaths due to lack of products. If, as Amalberti says, "the market economy prescribes the culture [of security] rather than the inverse," shouldn't it be the path to advance in prevention to leave the market economy?

In this text we seek two objectives: on the one hand, to deepen the understanding of the relationship between economic models and health, exploring the insidious mediations that, through neoliberal management devices, go from the financial to the subjectivity of the individuals, evidencing forms of resistance, but, above all, the limits of transformation, and on the other hand, the search for economic alternatives to the market, recognizing that capitalism, through the labor market, creates limitations to the full development of the abilities of salaried workers, including their physical and mental health. Here, the "democratization" of companies is the order of the day, in the most diverse forms: liberated companies, spaces of debate about work, institution of conflicts, and development of cooperation.

In order to develop our arguments, we first described the main changes of work in the twentieth century (see Sect. 3.2), noting how each regulatory mode produces specific forms of illness, including the most recent changes related to automation and services (see Sect. 3.3). The financialization of production and management brings new consequences to work (see Sect. 3.4), placing the need to look for new ways of regulating production-consumption-health without going through mercantile relations (see Sect. 3.5). We conclude the text with brief considerations about limits and possibilities of transformation (see Sect. 3.6).

3.2 Work in the Twentieth Century: Classic Forms of Work Organization and Their Limits

The twentieth century was characterized by major changes in labor. In the beginning of the century, the consolidation of the large industry as the main locus for the production of commodities, the adoption of the division of labor as proposed by Adam Smith, and the evolution of machinery constitute the context in which new forms of work organization emerged at that time. Two main work organization models were systematized and became dominant models: Taylorism and Fordism. For both models, work was considered as likely to be completely objectified; thus, the first principle of so-called scientific management, proposed by Frederick Taylor, recommended that work should be described, analyzed, and reorganized, using time and motion techniques, in order to reduce task duration. Tasks should then be prescribed by the technostructure (mainly engineers and technicians) and workers should follow prescriptions, which allowed for further control by managers, since, in the Tayloristic logic, once task duration (and then production process lead time) was prescribed, it was possible to plan production quantities for a given period. The goal of reduced task time was related to the competitive conditions for capitalism at that time, in a context of immature markets where profits could be generated by producing and selling standardized commodities. This was clear within Fordism and mass production. In the assembly line, labor was even more divided as a way to assure increased production volumes.

Therefore, in classic work organization, labor was considered to be equal to worker's gesture associated to machines or tools. Variability was considered to be rare, and the assumption of perfect rationalism of science, and as a consequence of scientific management, supported the view that prescribed work (or *task*, as French ergonomics conceptualizes it) could and should be equal to actual performance (or *activity*). Prioritized coordination mechanisms (Mintzberg 1993) were direct supervision and standardization of work processes. Mutual adjustment and lateral coordination were not considered as adequate coordination mechanisms, following the assumption of perfect rationalism.

The limits of Taylorism and Fordism have been widely discussed. Occupational health issues, such as occupational diseases, were evident due to work intensification and to separation between conception and control, on one side, and execution in the other: workers' room for maneuver in their activity could be strongly limited, notably in assembly lines. In this respect, Taylorism and Fordism, as models of organization, are a "reinvention of manufacturing in the twentieth century" (Moraes Neto 1989), resuming historically backward forms of work and organization in economic sectors artisanally organized, such as car production in small workshops before Ford created its big industry. Whether in the manufacture of the seventeenth and eighteenth centuries or in reinvented manufacture, the organizing principles are the same, essentially the division of labor into repetitive tasks and coercive domination by the factory hierarchy, to make docile rebels from individuals who do not accept as natural the submission to the conditions of factory work. This form of organization produces specific forms of sickness. Its mutilating character was portrayed by factory inspectors in England and authors of the time, which Marx summarized as follows:

> The continuity of uniform work *destroys the tension and the impulse of the vital spirits*, which find their recreation and stimulation in the very change of activity (p. 270).
>
> (...) a worker, who performs his whole life a single simple operation, *transforms his whole body. into an automatic, unilateral organ of this operation* (p. 269)
>
> Manufacturing, in fact, produces the virtuosity of the detailed worker by reproducing within the workshop the naturally developed differentiation of crafts which he has already found in society, and by systematically pushing it to the extreme (p. 269).
>
> On the other hand, it reaches this social organization of the work process only by *welding the same worker to the same detail* (p. 273)
>
> The collective worker now possesses all the productive properties in the same degree of virtuosity and at the same time expends them in the most efficient way, employing all his organs, individualized in workers or groups of workers, solely for their specific functions. The *unilaterality and even imperfection of the partial worker* become their perfection as a

> member of the collective worker (for instance, *unilateral development of muscles, bone deformation*, etc.). The habit of exerting a unilateral function transforms it into its natural organ and grants safe action, whereas the connection of the global mechanism obliges it to operate with the regularity of a machine component (p. 275–6).

In the passages quoted, we highlight some instances that directly refer to the etiology of the various work-related musculoskeletal disorders (WMSDs) (unilateral worker, deformations of the body as a function of detail virtuosity, repetitiveness, etc.), which at the end of the twentieth century grew on an epidemic scale (Kuorinka et al. 1995). It is from them that derive all the attempts to reorganize work (the so-called new forms of work organization - NFWO) that take on a variety of forms – enrichment of functions (or tasks), semiautonomous groups, and participatory management – to restrain capital predatory tendencies in the use of the workforce. These forms of work organization are not recent and arose from questioning the organization based on Taylorist/Fordist principles.

But there were limits also from the capital's side. Increasing market saturation that followed the post-World War II economic growth period and the oil crisis in the 1970s led to new forms of competition and industry organization. Indeed, management literature pointed to the need of product differentiation (concerning, for instance, quality, variety, flexibility, innovation) in order to face new market constraints and the competition with Japanese companies in the USA. This could be done by way of marketing strategies but also by changing the production model, which meant transforming work organization.

The enrichment of job positions, inspired by the Human Relations movement, sought to counteract the Taylorist conception of the worker exclusively as *homo economicus*, attributing to him moral and affective values important for understanding his behavior at work. In fact, it does not constitute an alternative to Taylorism because it does not dispute the division between planning and execution, seeking only to approach management from workers through the negotiation of conflicts, the improvement of internal communication, and the recognition of informal leaderships. This movement does not recognize the existence of contradictions but serves to regulate conflicts, reorienting dissatisfaction to the objectives of the company. As the internalization of values and the communion of objectives between workers and management expands, it is possible to reduce hierarchical levels and coercive control and increase the scope of the performers' tasks, both vertically (assigning them inspection activities and of first-level maintenance) and horizontally (multitasking versatility, job rotation). Here the emphasis is on consensual decisions, as to align each behavior to the company's goal. Contrary opinions become beneficial as long as a consensus emerges within the confines of what is acceptable to the company. Dissent is pathological and therefore mechanisms of exclusion are created, sometimes assumed by the work colleagues themselves, depending on the degree of adhesion to the company or the competition among the workers.

Other forms of work organization challenge the assumption of separation between execution and control or conception, claiming for workers' engagement in the productive process. This means, for instance, that workers should be responsible for the execution but also for tasks such as quality control, equipment maintenance, and workplace cleaning. This configuration is typical of work organization inspired by Japanese companies, mainly automotive ones, as Toyota. The Toyota Production System (TPS), as it was called, also preconized teamwork at shop floor, introducing a logic of mutual adjustment in formal work organization. However, work conception still remained separated from execution, as tasks should be defined by engineering or management, even if workers could (and should) propose new production methods through suggestion systems. Indeed, *kaizen* or continuous improvement aiming at cost reduction is a duty for workers in the TPS, thus forcing engagement and introducing a new pressure to shop floor workers. On the other hand, there is also a possibility of competence development at shop floor and the beginning of recognizing the importance of tacit knowledge to build competitive advantage (Coriat 2000; Lazonick 2005).

Constant cost reduction, the leitmotiv of TPS, is but one of the possible modes of competition in the late twentieth-century capitalism. Other business and productive models emerged, relying in differentiation coming from focusing in aspects such as quality, variety, flexibility, innovation, and service within production systems (Boyer and Freyssenet 2000; Du Tertre 2012). In order to achieve these goals, so-called immaterial or symbolic work, that is, cognitive activity and the engagement of workers in intersubjective relationships, became more and more important. Besides, changes in the productive technological base led to more automatized production systems, even in service activities, which also transformed the nature of labor activity towards immaterial work, as in automatized processes where workers must supervise the functioning of equipment and interfere when variabilities arise. Unforeseen elements are then recognized as being part of production systems. In order to cope with these changes, work organization was modified so as to consider immaterial work as a way to achieve competitiveness and to accomplish quality on work life – a way of dealing with occupational health problems. The rationale in new forms of work organization conceives workers as capable of taking initiative and making decisions in face of unforeseen events; given these unforeseen events, to rationalize a production system does not mean to prescribe every aspect of the system, which would be an impossible task, but to recognize variabilities and design organizational structures to cope with them (Veltz and Zarifian 1992). Two key aspects to those new forms of work organization were autonomy and cooperation in workplace that were to a great extent synthesized in practices of teamwork, notably semiautonomous workgroups or self-managed teams.

Semiautonomous workgroups were the core of sociotechnical work organization model. Sociotechnical systems design theory views organizations as open systems, able to dynamically adapt themselves to the environment in an organic mode, in which social and technical aspects should be jointly designed and optimized. Based on principles such as minimal critical specification, multi-functionality, and multi-skilling, work should be designed around semiautonomous workgroups (Cherns 1987). Within these groups, in a context of minimal specification, workers could have autonomy to define their actions concerning production aspects such as operational procedures, sequence orders, quality aspects etc., as well as to cope with variabilities in general. More than that, they should be responsible for achieving production goals, which are often set through performance indicators that become the major form of control. Therefore, in sociotechnical systems design, there is a rationale of personal accounting and self-control (Eijnatten 1993). At the same time, there is a rationale of competence development at work; indeed, when coping with variability, taking initiative and making decisions, notably in a collective way, workers mobilize existing tacit and explicit knowledge while acquiring new tacit knowledge, thus building competences at work (Lave and Wenger 1991; Zarifian 1999). These competences are the foundations for competitive advantage in a dynamic environment, as they allow the production system to be flexible and innovative. In brief, new forms of work organization combine mutual adjustment, individual and team responsibility, and competence development. Depending on how these elements are arranged, contradictions may emerge in the workplace and bring along pressures, stress, accidents, and occupational diseases. The power of decision-making, however, does not go beyond the management of immediate tasks, and the most important decisions remain an exclusive attribute of management (production goals, quality, number of workers, new products, introduction of new processes and technologies, training needs, etc.). Obviously these aspects can be negotiated (as in the German or Swedish co-management), but they are objects of conflict and not consensual decisions.

This is also the limit of the various forms of "participatory management" that separates daily decisions from strategic decisions – those continue to be an asset of the top management. Total Quality Management (TQM), perhaps the form most common in Brazil, attributes to the workers the decisions about the "daily routine," allowing even that the operative procedures be defined with the participation of the workers. These forms of management, however, grant only limited autonomy to execution tasks without altering in depth the content of day-to-day work. Even the suggestions that are given by the workers and implemented do not constitute daily tasks but are occasional facts that little enrich their work, still of an essentially manual nature. More recently, the movement of "liberated companies" seems to ignore these historical experiences and move in the same limits, now even more pressured by goal-management systems (Coutrot 2018).

The next section will discuss in depth three situations where "immaterial work" became recognized as fundamental for companies to achieve competitive advantage: the automation of production systems, the emergence of the "service relation" as a competitive distinction for companies, and innovation activities.

3.3 "Immaterial" Work: Automation, Innovation, Services

What has been so far written about the NFWO has left out automation, which promises to shift effort to the technical system and relieve operators of heavy, risky, or unhealthy work. We will see below that this new productive rationality has deep reflections on the subjectivity of workers and on the organization of work, depending on how one understands the need for responsibility and personal involvement in decision-making, how to deal with the possibility of error, and how the global working time is divided between time for action and time for reflection on action. Objective necessities put forward by how the automated system works and how it should be managed require from operators the ability of understanding unexpected and unique events that happen outside routine operational patterns and a new form of communicative and reflexive rationality. Historically, the continuous process industry (CPI) has been the "laboratory" of automation, which only recently, with robotics, has begun to penetrate the industry as a whole.

3.3.1 Mental Suffering in the CPI

We know that work in several subsections of CPI is intrinsically and potentially pathogenic, depending on the characteristics of the process industry, in particular due to the risks of explosions and exposure to chemical agents whose health effects are still unknown. These may be the most obvious aspects, revealed by research conducted at nuclear plants and chemical industries. Less commonly known is the daily wear and tear caused by the temporal requirements related to the uninterrupted flow and the uncertainties arising from the complexity of the installations. This form of wear (commonly attributed to stress) can hardly be manifested as specific occupational pathologies but mainly as psychical suffering which may also have nonspecific physiological effects (hypertension, sleep problems, etc.).

In CPI, the unequal distribution of free time to reflect about the process at hand is one of the main obstacles to the institution of a new productive rationality. One of the fundamental contradictions is manifested when one intends to define the space for autonomy of the operators. Formally, the spirit of initiative and personal responsibility are valued and expressly encouraged; in practice, there is little room for these good intentions to materialize. The pathological relation to the error is revealing of the psychic suffering that this situation causes, leading the workers to assume an infantile relationship with management. In a hierarchical organization, there is no room for the free development of the workers' personality, which must be accompanied by an expansion of scope and decision-making power. In several situations, workers are forced to submit to hierarchical power, without being intimately convinced that the orders being received are substantiated. Due to the peculiar characteristics of the production process in the CPI, the knowledge needed to control the process is strongly contextualized. An intervention in the process, the correction of a variable, the underlying diagnosis, and the way of changing the parameters are all activities that imply a specific way of relating to work situations, deeply rooted in the operators' previous experience and the here-and-now experience of the process. Each working day, "to put oneself in situation" requires time and availability on the part of the operator, as can be seen in shifts handover. It is frequent that, because bosses are only tangentially in contact with the process, they give (sometimes imperative) guidelines on how to deal with a given incident or problem. Most of the time, these directions or orders are unnecessary, either because the workers already know how to behave in that situation or because they are inadequate. Sometimes, depending on their bosses' insistence, operators do what they're asked to, knowing that the problem will not be solved and will end up causing them a greater workload because the process becomes more unstable. It is most common that workers act in a way that they themselves consider more correct, assuming a double risk: If it works out, the merit is the boss'; if it goes wrong, it is the fault of the operator because he disobeyed the guidelines. The paradox in this situation is that engineers and the immediate leadership are also dissatisfied with the behavior of the operators, claiming that they show little initiative, often resorting to superiors to solve "small" problems that they themselves could solve. In addition to the cognitive gaps (mosaic knowledge about the process), part of these difficulties is related to the division of responsibilities and the way in which hierarchical institutions deal with "errors," that is, in terms of guilt.

The infantilizing aspect of this relationship of subordination is that, in order to escape from the unilaterally assigned responsibility, the workers seek to behave the way their boss wants, acting like the children submissive to their parents or as students before the teachers. Even when they must decide on their own, they wonder what the boss' decision might be, or what he/she would like. In these conditions there is no room for the development of a mature personality, as workers are afraid to make decisions and publicly take responsibility for their actions. Obviously such an attitude is impossible as long as the power relations are maintained and any errors are used as a tool for selecting workers instead of serving as material for reflection and personal growth. Workers who more frequently and openly disrespect received orders they judge unfounded; therefore those who show an autonomous personality capable of initiative are considered undisciplined and the first to be dismissed when there is a reduction of the workforce. Due to the impossibility of relying on technical competence criteria, the hierarchy ends up clinging to the external signs of behavior, which, however, induce the operators to adopt an exact opposite behavior to what is theoretically sought: dependent and not autonomous.

3.3.2 The Pressure of Innovation

In the beginning of the twenty-first century, innovation arose as a strong element to attain competitive advantage, as it transforms opportunities into new ideas and put them in practice (Tidd and Bessant 2009). Innovation is also largely recognized as an important factor for socioeconomic development. Indeed, organization for innovation is a matter of management and labor studies but also a matter of public policy.

Production systems may be defined as composed by three spheres, which are interrelated (Veltz 2000): the sphere of activities of contact with the customers or users (for instance, marketing); the sphere of activities of exploitation of productive process, maintenance, planning, and optimization of current processes; and the sphere of activities of conception of new products, new services, and new processes.

While the former spheres deal with current or short-term-related activities, the latter deals with future or long-term-related activities. Contact and exploitation (production and support) activities refer to the present, and conception activities refer to the future. As a consequence, even if there is increasing space within contact and exploitation activities for abstraction and cognition, in conception activities abstraction is the rule and cognition is the very nature of work. The work of innovation is always related to a nonexistent object, a virtual object which is in constant transformation (Veltz 2000). Elements such as uncertainty and risk are inherent to innovation.

Innovation activities are also cumulative and collective (Lazonick and Mazzucato 2013). Innovation is about learning; that is, in order to innovate, one relies on what has already been learned while developing new knowledge; innovation is built over existing tacit and explicit knowledge and creates knowledge. As innovation becomes more complex, due to new technologies and market characteristics, different actors (including customers, users, competitors, public agencies, etc.) must contribute to the process, aiming at an exchange of knowledge to improve the learning process; hence innovation is collective. In this sense, organization for innovation should privilege cooperation and communication among actors with different specializations. Another characteristic of the innovation process is its performative nature; that is, it is an indeterminate and open process in which success or failure are not steady and well-defined concepts and in which the context for action is not given but constructed by the actors involved (Garud et al. 2017). Thus, innovation process cannot be completely predicted, and positivist approaches for work organization and management of innovation can hardly succeed.

Therefore, work organization for innovation should take into account these aspects – risk, uncertainty and collective and cumulative learning, as well as performativity. When organizing for innovation, literature suggests that elements of Taylorism and bureaucratic structures should be avoided; conversely, mutual adjustment through extensive communication across functions and hierarchical levels, teamwork, high level of involvement of workers in innovation processes (through, for instance, incremental innovation or continuous improvement/*kaizen* programs in shop floor), continuous efforts for competence development, and a reduced number of hierarchical levels (Dougherty 2008; Tidd and Bessant 2009) should be stimulated. Smaller degrees of specification are also considered to foster creativity, even if some constraints may contribute to innovation as they may drive the search for new solutions in order to overcome these constraints (Dougherty 2008; Lampel et al. 2014; Rosso 2014). In innovative project management, for example, methods such as "agile development" propose a sort of "minimal critical specification" to workers, whose work is developed within iterative cycles over the project (Hodgson and Briand 2013). Needless to say, the immaterial nature of the core of innovation activity turns rigid working hours useless, since part of the process of coming up with new ideas and problem-solving may be achieved at any moment, even outside work hours.

In broad, work organization becomes more flexible. Flexibility is also observed in work hours and, in many instances, in flexible labor contracts. In the software industry, for instance, casual contracts and self-employment have increased even after the boom of the so-called new economy (Bergvall-Kareborn and Howcroft 2013). This may create a paradox for the competence development in the long term, i.e., cumulative learning, which is needed to innovative activities. A similar problem arises with the notion of entrepreneurship as a major source of innovation. Even if individual creativity and entrepreneurship may constitute important elements for the innovation process, it is far from being sufficient; due to the collective and cumulative nature of innovation, organizational elements are at least as important, as they foster tacit knowledge transfer and allow individual creativity to be converted in innovation (Bharadwaj and Menon 2000). For instance, as innovative companies such as Google and 3M are known for encouraging their employees to develop their ideas in semiautonomous ways, this is achieved through organizational arrangements such as formally conceding time within the labor journey to develop such "pet innovation projects" (20% of the – formal – labor journey in the case of Google and 15% in the case of 3M), flat structures, teamwork, open communication, and rewards for innovation (Garud et al. 2011; Savoia and Copeland 2011). Additionally, small companies responsible for high-tech product, service, or process innovations often rely on networks with universities, research centers, or large companies in order to develop their new products, services, or processes (Lazonick 2005; Lazonick and Mazzucato 2013). Thus, in order to succeed in sustainable innovation processes, companies must balance individual initiative with collective activity and long-term competence development, as well as tacit knowledge management and transfer issues. Whenever such balance is not present and demands on innovation outcomes exist, there may be new pressures to workers related to the need of self-development in order to face job insecurity and to guarantee their own "employability."

Furthermore, as in the case of sociotechnical systems design, more flexible and autonomous workplaces do not mean that control structures are absent. In recognition to the fact that innovation activity is to a great extent unpredictable – thus making control through standardization of procedures almost impossible – innovation labor is often controlled through goals: for example, the number of innovative products/services/processes launched within a given period, the time to market of innovation processes, and respect to innovation budgets. These performance indicators are based on volume and time, typical Taylorist-Fordist objectives, and are frequently monitored by means of information technology (for instance, it is possible to monitor in real-time workers' interaction with a given software and therefore to "presume" work time and idleness). Nevertheless, due to the nature of the innovation process, these performance indicators may not fit. Results may come only in the long run, and not infrequently an unsuccessful innovation at a given time may become new products, processes, or services in the future. This mismatch between the nature of the activity and its mode of control may be a source of major occupational health issues (see, for instance, Waters (2017) for an analysis of suicides at Renault's French engineering departments).

Another effect, often unrecognized, is the functional use of drugs due to work requirements, which affects professionals with a high physical load (construction workers, sugar cane cutters, etc.), high attention levels (drivers, pilots, etc.), and also creative professionals (artists, marketing, etc.), each one according to different etiological processes. But they have in common the fact that drug is initially a functional tool to answer to work demands and, little by little, it becomes the problem to be solved when chemical or psychological dependence is installed and affects negatively both the worker's health and productivity (Lima 2010).

3.4 Financialization of Productive Systems and Its Consequences to Work

Another source of changes for work and work organization is the process of financialization of productive systems. Financialization means "(...) the web of interrelated processes – economic, political, social, technological, cultural etc. – through which finance has extended its influence beyond the marketplace and into other realms of social life" (Van der Zwan 2014, p.101). This phenomenon has been understood from three approaches: financialization as a regime of accumulation (successor of the Fordist regime); financialization of the modern corporation, that is, the subsumption of the organization and management of corporations to the logics of shareholder value maximization (SVM); and financialization of everyday life, that is, finance becomes a "governmentality" that takes root and, in a way, disciplines the practices of daily life (Van der Zwan 2014). These three approaches are interrelated, reinforcing each other. In this sense, financialization of productive systems means the process of subordinating an industrial, or productive, rationale to a financial rationale; decisions within productive systems that once followed a productive rationale now follow a financial one.

In the process of financialization, under the SVM logic, the company is seen as a "nexus of contracts" (Fligstein 2001), or a large "portfolio of investments," in which adding value through production is only one of the modes of capital valuation, which will be continuously evaluated in relation to other investment possibilities. This means that competition for productive activities does not take place between equal agents (other productive systems) but between different ones (productive systems but also financial systems). In practice, it is not enough for production, for example, to achieve excellence in the criteria traditionally evaluated in the management of operations – costs, quality, speed, reliability, and innovation – if this does not mean a return to the shareholder at levels equal to or greater than the shareholder would obtain in any other investment, considering the liquidity of the capital. In addition, a scenario of financialization means that the valuation of the activities carried out internally to the organizations is accomplished based on a financial rationality, which favors liquidity and the short term (Chiapello 2015).

As a consequence of financialization, companies adopt a strategy of "downsizing and distributing" their profits to shareholders, instead of "retaining and reinvesting" them in their productive systems (Lazonick and O'Sullivan 2000; Froud et al. 2006).

Downsizing movements, which include not only outsourcing but also flexible labor contracts, fit the rationale of liquidity and increase return on assets, but they also constitute a discursive construct (Froud et al. 2006; Van der Zwan 2014) as the simple announcement of these movements has an effect on company shares value, independent of real, long-term effects on company's profits. This reflects the performative character of financial valuation. Downsizing also means permanent cost cutting across all hierarchical levels and organizational functions. On administrative functions and shop floor level, cost and quality performance is a matter of personal accounting; that is, workers are responsible for achieving predefined goals in order to "create" and not "destroy value" to shareholders, mainly with the aid of performance indicators (Ezzamel et al. 2008; Cushen 2013). On the other hand, as a way of encouraging managers and labor efforts towards maximizing shareholder value, it is not uncommon that not only executive but also labor compensation is at least partly constituted of stock options. Therefore workers must continuously deliver high performance in order to accomplish constantly tougher goals and to maximize shareholder value, in this sense guaranteeing their own revenues (Ezzamel et al. 2008; Cushen 2013; Van der Zwan 2014). In addition, achieving these goals does not guarantee satisfying other stakeholders such as clients, suppliers, and other institutions, which means workers may need to manage these paradoxical injunctions in a daily basis (Gaulejac 2005). Damaging effects to workers' health are evident in this scenario.

Another important aspect is the claim for workers' engagement in order to generate value, be it to external stakeholders or shareholders, while companies themselves adopt flexible labor contracts and/or downsizing policies.

The specific efficacy of neoliberal management in its contemporary form, which serves as a mediator between financial logic and workers' control, is to institute individualism in the subjectivity of each worker. For this, it is necessary to break collectives and cooperative relationships (Linhart 2009; Dejours 2015); however, in order to institute a "new way of the world" (Dardot and Laval 2009), it is also necessary to penetrate and conform subjectivity in its own subjective processes, as well as the values assumed by every individual. One of the efficient elements of the new mechanisms of domination is that the intersubjective processes of social recognition that we need and that are positive for our individual development and our mental health are appropriated by the company in terms of quantifiable evaluation of individual performance (Vidaillet 2013), which makes management a "social disease" and expands the different forms of harassment (Gaulejac 2005; Gaulejac and Hanique 2015). This, however, is not done without resistance to paradoxical injunctions, more or less anesthetized by mechanisms of rationalization (Dujarier 2015), which can lead to severe decompensations such as burnout and panic syndrome.

When we study management methods that individualize performance and associate them with financial compensation, from the worker to the CEO, it is evident how the collective and the cooperation implode and the losses that this represents for the company. Revealing the irrationality of the new neoliberal reason is not enough to promote change. A contradictory system must, necessarily, naturalize its intrinsic irrationalities. The resulting human and material wastes are considered only as illegitimate incidents or resistances that will eventually be overcome. Maximum performance must be obtained by maximum competition, whose model is the professional athlete. Instituting conflict, and not the conflict debate, as Clot and Gollac (2014) want, is the goal pursued by management devices, even though that may seem absurd to everyone who knows the real work. As one French CEO puts it: "Cooperation is a mask under which individuals are unaccountable!" (quoted in Dejours 2015). The problem with that, which seems to respond to the developmental desires of each one, is that the objective of the professional athlete is not to stay healthy but to obtain the maximum performance: according to this, his body is put to work in such a way that brings sequels manifested outside his professional life, which is usually short and intense. Here, too, drugs and doping are widespread resources. It is not by chance that cooperation becomes the target of the new neoliberal reason but also the main resource to create alternatives that associate production and health.

3.5 It Is Possible to Come Out of the Market: Towards a New Economy and Regulation of Production-Consumption-Health

If direct interventions for labor transformation and social regulation, in their purpose of domesticating capital, always seem to run into insurmountable limits,[1] would there be forms of resistance and alternatives of health production even from within them? How can one get out of market domination and financialization to build a job that is a source of health and personal fulfillment?

In the context of capitalist enterprises, several experiments, still in initial stages, appear here and there, still without setting a trend of important social transformation. The "liberated companies" are bold in their objectives but conservative in their relationship with the market, believing that it is possible to promote meaningful work without transforming the relations of property engineering or financialized management. The actual experiences show the setbacks that this proposal of change from within, rather than radical, can lead into (Coutrot 2018).

Another broad movement is that of alternative economies, called solidarity or social economy, which attempts to build self-managed organizations, either by starting from scratch or by recovering failed capitalist enterprises. This movement gains amplitude particularly in times of strong economical crisis, with or without the support of leftist governments; however, these experiences seem to encounter barriers

[1]About the difference between "barriers" and "limits," we refer here to Marx (The Capital, more specifically the preparatory writings, the *Grundrisse*). In its development process, a social entification comes across barriers that are overcome but is always moving within limits that define it as a specific entity: those cannot be overcome unless there is a qualitative transformation of the very entity, giving rise to another entification (see Lima et al. 2015).

that constrain them to a local, small or medium scale, which limit their power of social transformation.[2]

Not surprisingly, among the initiatives taken within the solidarity economy and other actions that seek to develop alternative economies, one of the most successful has been agroecology, which has as one of its main guidelines the progressive elimination of pesticides, bringing health to producers and consumers. Several experiences of agroecology, such as that of MST (Movimento dos Trabalhadores Sem Terra – Landless Workers' Movement) in Brazil, the short circuits of food production in Europe, are exemplary in this regard. However, while successful initiatives as social movements and in recovering human dignity, they still face economic constraints that prevent them from obtaining economic gains and working conditions that are consistent with the relevance of their social and environmental work.

Despite the remarkable cases of success and the recognition that the economic determinations of industrial capitalism are a common point in all these critical approaches and institutional attempts to regulate and search for alternatives to the market economy, the same emphasis is not given on the construction of alternatives. These initiatives are often driven by genuine interests of social transformation, but their historical impotence may reveal not only the excessive power of capital and the strength of economic limits but also insufficiencies of theoretical frameworks and intervention strategies. It is worth remembering that, since its creation, ILO's main objective is to protect workers through the establishment of an international standard of competition, avoiding regional economic advantages based on social dumping.

In the work science field (ergonomics, psychology, ergology), the question of effectiveness and limits of the effective transformation of the current reality is also being questioned. The group coordinated by Rodolfo Vilela and Ildeberto Almeida saw in the proposal of intervention by the Change Laboratory (CL), proposed by Engeström, the additional effectiveness that would be lacking in ergonomics; it organizes the intervention, aimed directly to change, in spaces of debate between actors working on contradictions of work systems. Marçal Jackson, on the other hand, instigates ergonomics to adopt a "public" vocation, inspired by the public sociology of Burawoy (2014). Faced with these different social and institutional initiatives, the economy of functionality and cooperation (EFC) proposes a direction that also has to be considered. Born in France from practical interventions in socially and economically depressed territories after the end of coal-based economic activity (mining and steelmaking), EFC develops projects combining ergonomics, work psychodynamics, and economics, proposing an alternative to integrate work, health, and economic efficiency. The differential of the EFC to other

[2] One of the most notable counterexamples is Cecosesola, which operates in Barquisimeto, in the state of Lara, Venezuela, which manages large-scale funerary activities, supermarkets, outdoor fairs, and healthcare providers (hospitals and clinics). Unlike other traditional examples in the social economy, Cecosesola was able to develop and maintain horizontal management and has been operating for more than 50 years without funding from the Venezuelan state. See Cecosesola (2003, 2007, 2009).

approaches is precisely the emphasis on the development of an economic model integrated with work and based in cooperation, excluding market-determined social relations.

The EFC, like other critical approaches, recognizes the determinations and economic limits imposed by industrial capitalism on the development of human activity and, therefore, proposes as an alternative the development of economic enterprises within and by means of cooperative ecosystems of innovation capable of articulating production, work, and consumption (reproduction) in an enriching way for all social actors. In these alternative production systems, the work recovers its centrality, backed by immaterial values of knowledge (competencies), pertinence, trust, and health. Cooperation develops in all directions: horizontal, between peers; vertical, between management and operational teams; and cross sectional, between service providers and users. If everyone talks about work and health, the differential of this approach is precisely its articulation around alternative economic models that give sustainability to the strategies of change. Thus, economic, social, and environmental sustainability are no longer goals connected by the will only: they are treated as integrated and coherent projects in a long-term perspective. But like the other approaches, the problem arises of moving on to transformation on a social scale, and successful experiences are still limited to local spaces.

3.6 Final Remarks

This chapter discussed some characteristics of work and work organization in the twenty-first-century capitalism. Changes in society, technology, and markets implied changes in production systems towards an increased appreciation of immaterial work, which was undeniably followed by changes in methods of labor coordination and control and, to a certain extent, new impacts for occupational health.

It is interesting to note that many of the recent changes in work activity converge to a form of control through performance indicators. Indeed, if standardization of procedures has always been a poor proxy for labor activity, with the recognition of immaterial work as the main source of profitability, this proxy becomes totally inappropriate. Thus, predefining tasks as a means of labor control is rejected and immaterial work is perceived, at a first glance, as a possibility of more autonomous activities. However, performance indicators may generate new pressures towards labor, notably when objectives are defined by management alone without being negotiated with workers. This may imply work intensification and introduce "management by terror": as managers look to labor outcomes by checking achieved goals, they simultaneously overlook labor process and its procedures but also workers' efforts in order to accomplish these objectives. In other words, labor activity becomes opaque as management assesses labor in a "binary" mode, evaluating work positively if goals are attained and negatively otherwise, not taking into account any effort the worker may have made.

In practical terms, the question of transforming this reality divides the strategies among those who believe that it is possible to promote transformations from within: from the voluntarist and socially naive proposal of the liberated companies to the various forms of instituting internal spaces to debate work (Falzon 2014; Rocha et al. 2015), recognizing contradictions and conflicts as drivers of organizational, collective, and individual development (Engestrom 1987; Virkkunen and Newnham 2013; Clot and Gollac 2014). On the other hand, other strategies and approaches try to build alternatives in the pores or in spaces outside of the markets submitted to the financialization or the value chain dominated by large companies, giving up action in large financially owned companies, which reduce the spaces of authentic debate and rationality based on use value: the solidarity economy and EFC. Both strategies of transformation run into barriers that appear as limits: the managerial power of boards of directors that remain deaf to the debates at the lower levels or the power of the great transnational economy that dominates world markets and the great mass of workers and consumers, appropriating most of the value created. If strategies for action start from different social spaces, some essential values are shared: a social economy based on use value, more on services actually rendered than on material goods and monetary values; an organization based on cooperation in all directions, horizontal, vertical (with the very possibility of radical self-management), and transversal; and the possibility of creative and individually enriching work as a condition for the promotion of physical and mental health. Production, consumption, and health finally reconciled. In both cases, the limits are now set by the large financially driven capital that maybe will not easily give in, either eroded from within or dispossessed from outside. Although history is always relatively unpredictable as to when ruptures occur, the social powers to promote broad social transformations are not yet recognizable. In the meantime, we remain with these limited experiences, whose obstacles must be rigorously analyzed in terms of boundaries and barriers if we are not to return to the utopian voluntarism of good souls who want to transform the world without the objective conditions for it.

References

Amalberti, R. (2016). *Gestão da segurança: teorias e práticas sobre as decisões e soluções de compromisso necessárias*. Botucatu: FMP-UNEP.

Augusto, L. G. S., Carneiro. F. F., Pignati, W., Rigotto, R. M., Friedrich, K., Faria, N. M.X., et al. (2012). *Dossiê ABRASCO – Um alerta sobre os impactos dos agrotóxicos na saúde*. Rio de Janeiro: ABRASCO; 2ª Parte.

Bergvall-Kåreborn, B., & Howcroft, D. (2013). 'The future's bright, the future's mobile': A study of Apple and Google mobile application developers. *Work, Employment and Society, 27*(6), 964–981. https://doi.org/10.1177/0950017012474709.

Bharadwaj, S., & Menon, A. (2000). Making innovation happen in organizations: Individual creativity mechanisms, organizational creativity mechanisms or both? *Journal of Product Innovation Management, 17*(6), 424–434. https://doi.org/10.1016/S0737-6782(00)00057-6.

Boyer, R., & Freyssenet, M. (2000). *Les modèles productifs*. Paris: La Découverte.

Burawoy, M. (2014). *Marxismo sociológico*. Alameda: São Paulo.

Carneiro, F. F., Pignati, W., Rigotto, R. M., Augusto, L. G. S., Rizzolo, A., Faria, N. M. X., et al. (2012). *Dossiê ABRASCO – Um alerta sobre os impactos dos agrotóxicos na saúde*. Parte 1 - Agrotóxicos, segurança alimentar e nutricional e saúde. Rio de Janeiro: ABRASCO.

Carneiro, F. F., Augusto, L. G. S., Rigotto, R. M., Friedrich, K., & Búrigo, A. C. (2015). *Dossiê ABRASCO: Um alerta sobre os impactos dos agrotóxicos na saúde*. Rio de Janeiro: EPSJV; São Paulo: Expressão Popular.

Cavechini, C., & Barros, C. J. (2011). *Carne e osso. O trabalho em frigoríficos* [documentary]. Repórter Brasil.

Cecosesola. (2003). *Buscando una convivenciaarmónica*. Barquisimento: Digesa Lara.

Cecosesola. (2007). *Construyendo aqui y ahorael mundo que queremos*. Barquisimento: Digesa Lara.

Cecosesola. (2009). *Hacia un cérebro colectivo? De reuniones... a espacios de encuentro*. Barquisimento: Editora C.A.

Cherns, A. (1987). Principles of sociotechnical design revisited. *Human Relations; Studies Towards the Integration of the Social Sciences, 40*(3), 153–161.

Chiapello, E. (2015). Financialisation of valuation. *Human Studies, 38*(1), 13–35. https://doi.org/10.1007/s10746-014-9337-x.

Clot, Y., & Gollac, M. (2014). *Le travail peut-il devenir supportable?* Paris: Armand Colin.

Coriat, B. (2000). The 'abominable Ohno production system'. Competences, monitoring, and routines in Japanese production systems. In G. Dosi, R. Nelson, & S. Winter (Eds.), *The nature and dynamics of organizational capabilities*. Oxford: Oxford University Press.

Coutrot, T. (2018). *Libérer le travail*. Paris: Seuil.

Cushen, J. (2013). Financialization in the workplace: Hegemonic narratives, performative interventions and the angry knowledge worker. *Accounting, Organizations and Society, 38*(4), 314–331. https://doi.org/10.1016/j.aos.2013.06.001.

Dardot, P., & Laval, C. (2009). La nouvelle raison du monde. In *Essai sur la société néoliberale*. Paris: La Decouverte.

Dejours, C. (2015). *Le choix. Souffrir au travail n'est pas une fatalité*. Paris: Bayard.

Diniz, E. P. H., Assunção, A. A., & Lima, F. P. A. (2005). Prevenção de acidentes: o reconhecimento das estratégias operatórias dos motociclistas profissionais como base para a negociação de acordo coletivo. *Ciência & Saúde Coletiva, 10*(4), 905–916. https://doi.org/10.1590/S1413-81232005000400014.

Diniz, E. P. H., Pinheiro, L. C., & Proietti, F. A. (2015). Quando e onde se acidentam e morrem os motociclistas em Belo Horizonte, Minas Gerais, Brasil. *Cadernos de Saúde Pública, 31*(12), 2621–2634. https://doi.org/10.1590/0102-311X00112814.

Dougherty, D. (2008). Bridging social constraint and social action to design organizations for innovation. *Organization Studies, 9*(3), 415–434. https://doi.org/10.1177/0170840607088021.

Du Tertre, C. (2012). Lean production et modèle de valeur. Une approche régulationniste par le travail. *Activités, 9*(2), 168–178. https://doi.org/10.4000/activites.462.

Dujarier, M.-A. (2015). *Le management desincarnée*. Paris: La Découverte.

Eijnatten, F. M. (1993). *The paradigm that changed the workplace*. Stockholm: Arbetslivscentrum - The Swedish Center for Working Life.

Engestrom, Y. (1987). *Learning by expanding*. Helsinki: Orienta-Konsultit.

Ezzamel, M., Willmott, H., & Worthington, F. (2008). Manufacturing shareholder value: The role of accounting in organizational transformation. *Accounting, Organizations and Society, 33*(2), 107–140. https://doi.org/10.1016/j.aos.2007.03.001.

Falzon, P. (2014). *Constructive ergonomics*. Boca Raton: CRC Press.

Fligstein, N. (2001). *The architecture of markets: An economic sociology of twenty-first-century capitalist societies*. Princeton: Princeton University Press.

Froud, J., Johal, S., Leaver, A., & Williams, K. (2006). *Financialization and strategy: Narrative and numbers*. London: Routledge.

Garud, R., Gehman, J., & Kumaraswamy, A. (2011). Complexity arrangements for sustained innovation: Lessons from 3M Corporation. *Organization Studies, 32*(6), 737–767. https://doi.org/10.1177/0170840611410810.

Garud, R., Gehman, J., Kumaraswamy, A., & Tuertscher, P. (2017). From the process of innovation to innovation as process. In A. Langley & H. Tsoukas (Eds.), *The SAGE handbook of process organization studies*. London: SAGE.

Gaulejac, V. (2005). *La société malade de la gestion*. Paris: Seuil.

Gaulejac, V., & Hanique, F. (2015). *Le capitalisme paradoxant*. Paris: Seuil.

Hodgson, D., & Briand, L. (2013). Controlling the uncontrollable: 'Agile' teams and illusions of autonomy in creative work. *Work, Employment and Society, 27*(2), 308–325. https://doi.org/10.1177/0950017012460315.

Kuorinka, I., Forcier, L., Hagberg, M., Silverstein, B., Wells, R., Smith, M., et al. (1995). *Lésions attribuables au travail repétitif*. Québec: Multimondes.

Lampel, J., Honig, B., & Drori, I. (2014). Organizational ingenuity: Concept, processes and strategies. *Organization Studies, 35*(4), 465–482. https://doi.org/10.1177/0170840614525321.

Laurell, A. C., & Noriega, M. (1989). *Processo de produção e saúde*. São Paulo: Hucitec.

Lave, J., & Wenger, E. (1991). *Situated learning: Legitimate peripheral participation*. Cambridge: Cambridge University Press.

Lazonick, W., & Mazzucato, M. (2013). The risk-reward nexus in the innovation-inequality relationship: Who takes the risks? Who gets the rewards? *Industrial and Corporate Change, 22*(4), 1093–1128. https://doi.org/10.1093/icc/dtt019.

Lazonick, W., & O'Sullivan, M. (2000). Maximizing shareholder value: A new ideology for corporate governance. *Economy and Society, 29*(1), 13–35. https://doi.org/10.1080/030851400360541.

Lazonick, W. (2005). The innovative firm. In J. Fagerberg & D. Mowery (Eds.), *The Oxford handbook of innovation* (pp. 29–55). New York: Oxford University Press.

Lima, M. E. A. (2010). Dependência química e trabalho: uso funcional e disfuncional de drogas nos contextos laborais. *Revista Brasileira de Saúde Ocupacional, 35*(122), 260–268. https://doi.org/10.1590/S0303-76572010000200008.

Lima, M. E. A., Araújo, J. N., & Lima, F. P. A. (1997). *LER: Determinantes econômicos e psicossociais*. Belo Horizonte: Health.

Lima, F. P. A., Rabelo, L. B. C., & CASTRO, M. L. G. L. (2015). *Conectando Saberes: Dispositivos sociais de prevenção de acidentes e doenças no trabalho*. Belo Horizonte: Fabrefactum.

Linhart, D. (2009). *Travailler sans les autres*. Paris: Seuil.

Marx, K. (1983). *O Capital*. São Paulo: Abril Cultural.

Mintzberg, H. (1993). *Structure in fives: Designing effective organizations*. Englewood Cliffs: Prentice-Hall.

Moraes Neto, B. (1989). *Marx, Taylor e Ford*. Brasiliense: São Paulo.

Rocha, R., Mollo, V., & Daniellou, F. (2015). Work debate spaces: A tool for developing a participatory safety management. *Applied Ergonomics, 46*(A), 107–114. https://doi.org/10.1016/j.apergo.2014.07.012.

Rosso, B. D. (2014). Creativity and constraints: Exploring the role of constraints in the creative processes of research and development teams. *Organization Studies, 35*(4), 551–585. https://doi.org/10.1177/0170840613517600.

Santos, M. C. O., Lima, F. P. A., Murta, A. E. P., & Motta, G. M. V. (2009). Desregulamentação do trabalho e desregulação da atividade: o caso da terceirização da limpeza urbana e o trabalho dos garis. *PRO, 19*(1), 202–213. https://doi.org/10.1590/S0103-65132009000100013.

Savoia, A., & Copeland, P. (2011). Entrepreneurial innovation at Google. *The Computer Journal, 44*(4), 56–61. https://doi.org/10.1109/MC.2011.62.

Silva, P. H. N. V. (2012). *Epidemiologia dos acidentes de trânsito com foco na mortalidadede motociclistas no estado de Pernambuco: uma exacerbação da violência social* [these]. Recife; Fundação Oswaldo Cruz.

Tidd, J., & Bessant, J. (2009). *Managing innovation*. Chichester: Wiley.

Van der Zwan, N. (2014). Making sense of financialization. *Socio-Economic Review, 12*(1), 99–129. https://doi.org/10.1093/ser/mwt020.
Veltz, P. (2000). *Le nouveau monde industriel*. Paris: PUF.
Veltz, P., & Zarifian, P. (1992). Modèle systémique et flexibilité. In *De Terssac G. Les nouvelles rationalisations de la production* (pp. 43–61). Toulouse: Cépaduès.
Vidaillet, B. (2013). *Évaluez-moi! Évaluation au travail: les ressorts d'une fascination*. Paris: Seuil.
Virkkunen, J., & Newnham, D. S. (2013). *The Change Laboratory: A tool for collaborative development of work and education*. Rotterdam: Sense Publishers.
Waters, S. (2017). Suicide voices: Testimonies of trauma in the French workplace. *Medical Humanities, 43*(1), 24–29. https://doi.org/10.1136/medhum-2016-011013.
WHO – World Health Organization. *Global Status Report on Road Safety 2015*. 2015. http://www.who.int/violence_injury_prevention/road_safety_status/2015/en/. Accessed 1 Oct 2017.
Zarifian, P. (1999). *Objectif compétence. Pour une nouvelle logique*. Paris: Editions Liaisons.

Part II
Change Laboratories

Chapter 4
Is it Possible to Run a Change Laboratory if the Company Does not Recognize its Demand as Legitimate?

Manoela Gomes Reis Lopes, Rodolfo Andrade de Gouveia Vilela, Ildeberto Muniz de Almeida, Silvana Zuccolotto, and Flora Maria Gomide Vezzá

Abstract In recent years, there have been 12 fatal work accidents at a metropolitan passenger railway company. The Labor Prosecution Office contacted the School of Public Health to carry out a research making a deep analysis to understand the accidents and implementation of changes in this company to prevent accidents. Considering the presence of a civil accident inquiry and that the demand did not come from the company, the negotiation process for implementing the research was delayed and resulted in difficulties during the ethnographic phase. The research is still under progress. This chapter presents partial results of the application of the methodology, highlighting difficulties and strategies adopted by researchers in order to overcome difficulties. One of the strategies was the realization of workshops at the end of the ethnographic phase, aiming at presenting Cultural-Historical Activity Theory (CHAT) concepts and consolidating cooperation between researchers and participants, a crucial point for the activity development.

M. G. R. Lopes (✉)
Department of Biological Sciences and Health, Federal University of Amapá, Macapá, AP, Brazil
e-mail: lopes.manoela@unifap.br

R. A. G. Vilela
Department of Environmental Health, School of Public Health, University of São Paulo, São Paulo, SP, Brazil

I. M. de Almeida · S. Zuccolotto
Department of Public Health, Botucatu School of Medicine, São Paulo State University "Júlio de Mesquita Filho", Botucatu, SP, Brazil

F. M. G. Vezzá
Sports and Health Promotion Executive Management, Industry Social Service - SESI, São Paulo, SP, Brazil

R. A. G. Vilela et al. (eds.), *Collaborative Development for the Prevention of Occupational Accidents and Diseases*, https://doi.org/10.1007/978-3-030-24420-0_4

4.1 Introduction

In Brazil and several countries, work accidents tend to be investigated in search of a culprit for its occurrence. Accidents are usually attributed to a few causes and often considered as resulting from unsafe act or unsafe condition. In an unsafe act, the injured worker is held responsible for the event's occurrence, which contributes to judicial protection of the companies in the accident (Jackson Filho et al. 2013) and hinders the prevention of new accidents, since there is no deep understanding of their latent causes.

The traditional approach creates a climate of insecurity and fear among company's employees. When an external expert comes into a company for accident analysis, this climate results in a greater time to overcome resistances and creates conditions of cooperation for it.

Besides, in many methodologies for accident analysis, including those that adopt a systemic approach, workers are involved as informers and not as participants in diagnosis and transformations (Osório et al. 2005). The Cultural-Historical Activity Theory (CHAT) and the Change Laboratory (CL) (Engeström et al. 1996; Virkkunen and Newnham 2013) can contribute to reduce these limitations of the accident analysis, since they presuppose that all actors at the organization participate and collaborate during the main steps, including the historical analysis of problems and their contradictions that will lead to creating solutions. One of the challenges in accident analysis participatory approaches is involving high hierarchy right from the beginning of the process in such a way as to facilitate trust and exchange, which are essential for cooperation and transformation.

Moreover, accident analysis methodologies are not based on learning theories, which hinders any effective organizational learning that could cause perennial changes favorable to the prevention of new accidents. In this way, we believe that CHAT and CL also can contribute.

As mentioned in Chap. 2, CL is a method of formative intervention based on the theory of expansive learning (Engeström 1987, 2015), which seeks the reconceptualization of the activity's object as well as agency development in the participants. It has been used as an alternative method for the study of organizational conditions in order to bring about change (Querol et al. 2011), but it is still new and little known in several countries especially in Brazil.

There are few studies that use CHAT for accident analysis (Nuutinen and Norros 2009; Yoon et al. 2016), and only one study was found using CL in this perspective (Lopes et al. 2018). When using CL as a tool in the study of work accidents in authoritarian contexts, the negotiation process is a moment that demands caution, since it is necessary to gain confidence of all levels of hierarchy to carry out the analysis and the research. Authoritarian work environments are common in Brazil, and it is difficult for workers to express themselves and discuss freely about the organization's problems, for fear of retaliation. Depending on the demand of the study and how negotiation takes place, it is believed that these barriers can continue to permeate the research.

Based on a situation experienced by the authors, this chapter proposes to answer the question: how can one involve the high hierarchy and other actors in the process of participatory construction of changes when, at the start, there is no recognition of the demand from top management?

The empirical case of a public metropolitan passenger railway company will be discussed. In recent period, several fatal accidents occurred involving workers of this company. This triggered a civil inquiry by the Labor Prosecution Office (MPT) which requested the School of Public Health of the University of São Paulo (Faculdade de Saúde Pública da Universidade de São Paulo – FSP/USP) to conduct a new analysis of the working situations that gave rise to the accidents. The MPT wanted to further understand organizational determinants that played a role in the production of accidents, as well as to identify solutions to improve safety, in a partnership that makes it possible to reconcile academic research with the aspirations of society.

Metropolitan passenger transport plays an important role for society, as it is responsible for the commute of thousands of people daily, whether for work or leisure. Accidents involving workers in this sector have bigger repercussions than the accident itself – they affect all these people, and companies have to innovate in organizational strategies to improve health and safety. They cannot do so without the involvement of the actors of the organization.

This chapter aims to describe the difficulties for accomplishing a CL when the demand is not recognized by the high hierarchy and possible strategies to enable the formative intervention.

4.2 Research Context and Negotiation Process

Railway activity is a complex activity, and this company, because of its size and diversity, brings together many different departments whose actions take place in various geographic spaces. It is the result of a merger of three different railway companies and, even after 22 years, its members still refer to the presence of the culture of companies X, Y, and Z inside the current company (company A). In addition, it is a hierarchical company, with a process of decisions and information that must follow the hierarchy, making it slow and also harming the return of experience.

Workers from different sectors reported that the company has been growing in number of users – more than 236% passengers' growth over a period of 18 years – but operates with budget constraints that impact over both expanding or maintaining the workforce and in the acquisition of materials and production resources. As a state-owned company, the acquisition of new materials and hiring of human resources depend on government approval and bidding processes. The need for public bidding often results in lengthy buying processes.

In the last 7 years, 9 work accidents at this Brazilian company by runover of trains resulted in 12 deaths and 2 surviving victims, who were either employees or outsourced workers.

As mentioned earlier, after this series of accidents, a civil inquiry was initiated by the MPT. The first demand from the MPT to our university team was an in-depth diagnosis of the events in a way that would allow to apply penalties to the company, in a traditional judiciary approach. The first learning process was presenting to the attorney the CL methodology as a new way for remodeling the work processes and so preventing similar events in future. The new perspective was collaboratively constructed as a new modality of penalization and company liability.

After this agreement between academics and MPT, a process of negotiation started between them and the railway company. The group of researchers appointed by the MPT was formed by professors from FSP/USP and the São Paulo State University Júlio de Mesquita Filho in Botucatu (UNESP-Botucatu) and technicians of the Jorge Duprat Figueiredo Foundation (Fundacentro).[1] Eight meetings involving representatives of the mentioned institutions were made over 10 months to negotiate what would be done.

As the company was resistant to the study, whose demand arose externally via MPT, the research group used, in order to favor the recognition of the need for the study, the strategy of double stimulation. A timeline was drawn up and presented, marking 41 accidents involving passenger trains between 1987 and 2012, totaling thousands of injuries and 270 deaths due to collisions and derailments. The presentation of the timeline was followed by a debate on what were the main facilities and difficulties of the company to achieve its objectives, defined in the institution's mission as the provision of public transport service with excellence and safety.

The strategy adopted generated discomfort and defensive positioning of the company representatives, since many of these accidents occurred before the merger of the three companies that formed the current railway company. The difficulty of integrating their different cultures into an integrated whole was recognized, but according to them the information brought as a mirror data did not reflect the situation at that time, since important changes had taken place: technological development, efforts for standardizing procedures, training of staff, and, according to them, improving organizational culture.

Since they did not recognize themselves in the presented mirror data, they invited the team of researchers to get to know the company. This started their insertion in the organization.

Choosing mirror data is an important step in the methodology, because it will interfere with the entire process. The participants should visualize their activity in the offered data. In this case, mirror data presented partially reflected a past reality,

[1] Fundacentro is a governmental occupational safety, health, and environment research institution created to produce and disseminate knowledge, incorporating these themes into the elaboration and management of public policies for "sustainable development with economic growth, promotion of social equity, and environment protection" (http://www.fundacentro.gov.br/)

different from the one experienced by the company at that time, and caused embarrassment among the company representatives, as follows:

> When discussing about the causes of the runover accidents by trains, one of the main company directors stated that they occurred because "the workers were doing wrong in a wrong place and in a wrong time." But the assertive was immediately contested by a safety engineer from the company that said: "Despite my long experience and awareness about following rules, once I almost was run over by a train."

This embarrassment during negotiation was an important event for our learning:

- To conduce a good negotiation, we have to choose mirror data carefully, as well as the best strategy for this critical moment.
- Despite the defensive posture of the top management, the voice of the safety engineering recognizing that accident risks are present in the work process shows the relevance of the topic for the company reliability.
- A negative image externally offered by the academics as a first stimuli is not a good strategy because this reinforces resistances and can provoke a distancing of these actors towards the research team, with repercussions in different moments of the research.

In addition, the existing civil inquiry, the tradition of a legal approach to workplace accidents, and the fact that most of the meetings took place at MPT facilities had a negative impact on two important aspects. On the one hand, this hampered the managers' recognition that there really was a problem. On the other hand, FSP/USP was seen as a representative partner of MPT which also hampered the process of accidents' reanalysis, because of the legal approach to workplace accidents.

After the team of researchers visited the organization and a few adjustment meetings, a research proposal was presented to the company, whose objective was an in-depth reanalysis of accidents and critical incidents for prevention of new ones. Three methods of analysis were foreseen: the Analysis and Prevention of Accidents' Model (Modelo de Análise e Prevenção de Acidentes do Trabalho – MAPA) (Almeida and Vilela 2010; Almeida et al. 2014), the Collective Work Analysis (Análise Coletiva do Trabalho – ACT) (Ferreira 1993), and the Change Laboratory (CL) (Engeström 2007; Engeström et al. 1996; Virkkunen and Newnham 2013). Based on this proposal, a cooperation agreement was reached, consisting of representatives from FSP/USP, UNESP-Botucatu, Fundacentro, MPT, and the railway company. This term established a research based in a participatory, collaborative format with internal actors of the company and researchers, in order to provide an environment of cooperation and organizational learning for accident prevention.

In February 2019, when we wrote this chapter, research was in progress with CL sessions under way. We hope to hold a collective, participatory analysis of work accidents by run over of trains. Ten weekly CL sessions will be held, bringing together workers from different company departments. We plan to apply the MAPA^{EX} (see Chap. 14), in order to promote a lasting organizational learning.

4.3 The Process of Collecting Ethnographic Data

The fieldwork began in January 2017 with five meetings, when company staff presented its different areas. Respecting ethical procedures and in order to guarantee anonymity, the railway company will be identified as company A in this chapter.

After this stage, ethnographic data collection was carried out through individual or collective interviews, observations of work activities, and documental analysis in 33 days of field trips. Of this total, 24 field trips were dedicated to watch workers' presentations on the different sectors of the company, to observe the activities, and also to conduct individual and/or collective interviews with 46 company employees, totaling approximately 29 hours of interview recordings. In addition, two trips were related to ACT, totaling approximately 5 hours of recordings.

The workers interviewed belong to different hierarchical levels and represented different departments of the company: operations, circulation, maintenance, work safety, and operational safety.

Observations of the work activities were made in conjunction by the research team and company representatives – middle management team members. Two sessions of Collective Work Analysis were held, attended by circulation and maintenance departments' workers and a union representative.

4.4 Main Difficulties of the Ethnographic Data Collection Phase

During the interviews made when collecting ethnographic data, it could be noted that participants had a certain fear of talking about their activities and difficulties experienced.

This defensive posture was credited to this study's demand – fatal runover work accidents at railway crossings and the civil inquiry. It is worth mentioning that MPT's involvement in negotiating the research with the company may also have contributed to a confusion about the roles of the FSP/USP and MPT institutions among the company's employees, as came up during interviews.

> [...] but how is one supposed to clash with the MPT? How to clash with you [FSP/USP researchers]? Because today you are MPT [...] For me you are MPT.

In addition, we believe that the choice of mirror data presented during the negotiation phase also contributed to this climate of mistrust regarding the researchers. In the ethnographic data collection, it came up that the timeline reflected the reality only partially and brought a negative image of the company, ignoring the efforts and strategies adopted over time to deal with the difficulties they had.

As it was mentioned previously, company A is structured in many hierarchic levels, and access to top management was difficult at this stage. The CL assumes a participatory and collaborative process involving different levels of the organization. It is important that top management is involved; otherwise the process and designed solutions will be not implemented.

CL is still new in Brazilian reality and uses little known concepts, such as activity system and network of activities systems, that could seem abstract for most of the company's workers. This new approach may have contributed, along with the other factors pointed out earlier, to a misunderstanding of what the research actually aimed at.

Lopes et al. (2018) conducted a CL in civil construction. They verified, when analyzing CL sessions data, that participants showed difficulty in grasping the concepts of activity system and contradictions. The idea of holding workshops on the concepts used in CHAT, MAPA, and CL was to minimize these difficulties in the railway company and help participants to make out the concepts and processes. In addition, the workshops would also be an opportunity to clarify research objectives and doubts among the participants, thus breaking resistances felt during ethnographic data collection.

4.5 Formative Workshops as a Strategy to Deal with Difficulties During Fieldwork

These workshops were planned after roughly a year of fieldwork and were intended for presenting MAPA and CL methods. It was a change in the original research project presented to the company, in which only one workshop was planned at the beginning of data collection stage. Nevertheless, sensing the embarrassment caused by the negotiation process, the research team delayed this workshop, waiting to have more data that could be used as a mirror of the company's current state.

In total, four workshops were made, gathering representatives from operation, circulation, work safety, and maintenance departments. These workshops were conducted on seven different days; 49 workers of different hierarchical levels took part in them; the first one was longer (16 hours) than the others (8 hours each), totaling 40 hours.

All the workshops were planned according to principles of the CHAT and CL, such as multi-voicedness and the double stimulation method, in which problems are used as a first stimulus and then as a second stimulus by being confronted to a conceptual model. The group was divided into subgroups to discuss MAPA and CL methods concepts, such as change analysis, barrier analysis, organizational determinants, bow-tie model, activity system and its historical evolution, contradictions, and mediations. The workshop ended with a collective discussion of the themes discussed in each subgroup.

As mentioned before, the initial proposal was only one 16-hour workshop (prior to CL sessions) to present MAPA and CL concepts to managers and workers from different departments: 17 workers attended to it, on 2 consecutive days approximately 1 year after the beginning of data collection.

On its first day, we started using, as a first stimulus, a real case of a worker that was beaten by a robot. Questioning was used to expand the understanding of usual work analysis, change analysis, barriers analysis, and this cases' organizational determinants. This group discussion was followed by a presentation of MAPA concepts – such as usual work, changes analysis, barriers analysis, and organizational determinants.

Still using the case of this worker beaten by a robot and a brief explanation of the activity system and contradictions concepts, we asked the participants to draw the activity system involved in this case, as well as its possible contradictions.

Searching to reinforce the concept of activity system, we then used a video showing the activity of mammoth hunting in prehistory. Participants were asked to identify the elements composing this activity system, pointing out its changes and innovations.

This was followed by another video that presented some airport's activities that involved interaction and shared objects among different activities systems. These stimuli were the raw material for participants to work on the concept of shared object.

On the second day, the activities proposed in the workshop were focused on the railway company. Participants were divided into subgroups according to the departments they worked in. We asked them to draw each department's activity system.

After presenting and discussing collectively this exercise, researchers made a presentation of CL concepts: unity of analysis, contradiction, expansive learning cycle, double stimulation, history and development, multi-voicedness, mediation, and role of researchers and actors during CL.

We got back to the concept of shared objects, focusing now on prevention of rail track runover accidents in the company. Two subgroups were formed, mixing people from different departments. They were asked to draw their activities systems, signaling their interactions and contradictions.

The discussion that followed the reunion of the subgroups was a very rich moment and brought up many elements that hadn't been seen during ethnographic data collection – for instance, a more detailed analysis of interfaces among the different departments.

The departments can be considered connected activities subsystems, in which the outcomes of each of them can be used as rules or instruments for the others; they constitute a network of activities systems (Virkkunen and Newnham 2013): there is interdependence and need for coordination among them. In addition, there are hypotheses of coordination difficulties between different departments, conflicts of goals that are set to them, their material and technical conditions, and about staffing levels for carrying out the activities.

Among the hypotheses of historical contradictions, we mention the change in technological quality and increasing demand of passengers and flow of trains (therefore a more complex object) versus a reduced workforce in a context of division of labor and rules that impose a reduced time for trains intervals and, as a result, a reduced time for maintenance activities.

An example of the analysis made by participants during the workshop
Workers explained that each department can be considered subsystems that have to interact in order to prevent accidents in the rail track (RT). Operational procedures (OP), designed by one specific department, are based on safety norms established by the occupational safety department. OPs become the rule for all the other departments: human resources, for instance, takes them as a basis for trainings it elaborates for all the company's departments. Access to RTs is the object of specific safety norms and OPs that have to be followed by everyone.

Besides those, in order to gains access to RT, maintenance workers have to ask for an access permission, which is discussed during fortnightly planning meetings with the operational strategy department. This meeting establishes RT access priorities that will have the least impact on trains circulation for the next 15 days. Once defined, the plan is handed to the Operational Control Center (OCC), responsible for train traffic. OCC, along with traction department, have to follow the plan to define which trains are going to run and their intervals: by doing so, it grants control and safety of all the workers that have to access RTs.

When a problem happens at the track, it is communicated to the maintenance information center, which decides if it is an emergency. If it is, a priority access permission is issued that can alter the fortnight planning. This has to be informed to OCC; otherwise the workers accessing RTs can become "invisible" to the system that looses track of them.

This example shows the interdependency among the company's departments and highlights the importance of coordination and communication, without which accidents can happen. The lack of coordination brings up a hypothesis of contradiction in the network of activity systems. Several objects should be tuned in by a script that defines actors and action sequences. When conflict arises between roles or concurring scripts, coordination problems happen (Virkkunen and Newnham 2013).

After this activity, yet another exercise was proposed. Based on the cycle of expansive learning, we asked the participants to plan CL sessions, working in subgroups. They should name all the sectors to be involved, as well as the number of people and of sessions and which activities should be done in them. However, there were few managers and some key departments that were not represented at this first workshop, and participants refused to do this exercise, since they didn't feel they had the authority to do the planning without higher managers, who have power to make the implementation of changes feasible. Thus, during the first workshop, participants proposed a second workshop gathering managers of all areas; and they proposed a reunion as well with the company's president to discuss and reinforce the theoretical aspects and systemic approach of the research and search for his commitment to implementing proposed changes in the future.

Chairmanship of this company is a trusted post of the state governor. Since an electoral process was to occur in that same year, it was not certain that he would continue on this position. Possibly by this reason, researchers couldn't be included in his agenda for the requested meeting up to the moment of writing this chapter.

We believe that this first workshop was a turning point for this CL intervention. The double stimulation strategy fostered participants' engagement, agency, and expansive learning during the workshop. They gradually grasped, in a practical way, the concepts related to the methods and understood the real objectives of the research. The participants pointed out that now they could understand the real purpose of the research and said they had been oriented to keep distance from the university team during the interviews, and some of them asked to be interviewed again as showed by the speeches below.

> We were oriented to keep distance from you when you had been at our department.
>
> If I knew that the research objective was this, my interview would be different. I would like to be interviewed again.

After the 1st workshop and based on the participants' feedback, we started the negotiation of a 2nd workshop with top management. The company's representatives proposed that 2nd and 3rd workshops were made, with the same content as the 1st one but shorter. These workshops were differently organized: two 4-hour meetings nearly a month apart. Even if the proposal was to have top managers participating, they could not fit the meetings in their agenda and designated middle-level managers to represent them. These three elements – the change in participants' hierarchical level, total time reduction, and a longer interval between them – had a negative impact on learning and made it difficult for participants to remember, on the second day, what had been discussed on the first meeting of each workshop. In addition, with reduced course load, participants also seemed to have greater difficulty in grasping the concepts when compared to participants in the 1st workshop. Even so, the workshops provided an important space for presenting and discussing theoretical concepts and seemed to contribute to decreasing participants' resistance to the research proposal.

In the end, a 4th workshop was organized for maintenance workers, whose absence in the 1st, 2nd, and 3rd workshops was noticed and questioned (both in interviews and during the workshops), since they are the ones who suffer the greatest impact and risk by performing their activities in a reduced period. This 4th workshop also lasted 8 hours; however, it was held in a single day.

At the end of the four workshops, our assessment was that participants grasped the concepts underlying the methods, because they had already started an interesting preliminary analysis of the metropolitan railway service activity system and its historical contradictions. Besides, workshops offered a space for breaking resistances, allowing to clarify what researchers were actually seeking: they became very participative and willing to collaborate in the next steps.

Unfortunately, the top management could not take part in the workshops. It is important to remember that this research has a participative approach to promote learning and presupposes participation and engagement of all the actors and hierarchical levels in order to establish a collective diagnosis of the problems which will be the basis for next phase – the visualization and implementation of changes.

During the workshops and interviews, it became clear that rank played a part in position and disposition of participants. High-ranking workers in general presented a legal and defensive view; field workers and intermediate management showed less resistance and more easily adhered to the research.

These workshops were an innovation in the conduction of a CL, as they hadn't been used before to introduce the method. They were a tool for reducing perceived resistance in the first stages of research. In addition, we expect that the assimilation of concepts will save time in CL sessions, allowing deeper analyses and richer data, influencing positively the construction of solutions.

All the stages described here will serve as a basis for CL sessions' participatory analysis and solution building. Data collected in the initial steps will not only assist in the elaboration of preliminary hypotheses but will also be used as mirror data, that is, data that reflect the activity to be analyzed during the CL sessions.

When they do start, participants will be encouraged to conduct collaborative analyses to understand historical sources of contradictions that may still affect current activity, thereby improving their work practices and building concrete and lasting innovations (Virkkunen and Newnham 2013) for improving quality, safety, and well-being at work as well as system's reliability.

4.6 Final Considerations

This study's demand, originated from a MPT decision of instituting civil inquiry and through MPT's mediation and also the strategy adopted by the researchers during the negotiation process, determined a climate of mistrust and fear among the participants. In addition, company representatives (top management) did not recognize, during negotiations, that a study was necessary, and this reverberated throughout ethnographic data collection.

The workshops on MAPA and CL methods played an important role in changing this climate and led participants to ask for another interview after understanding research objectives.

We could not have access to the top level management during these workshops. Different perspectives about the company are related to the hierarchical level.

Therefore, it would have been very important to have listened to their voice at this activity system, in particular because they are key persons to implement the proposed solutions.

It is still very common that organizational change or accident investigation methodologies place company members as informants and not as co-developers in the process of creating solutions. During the ethnographic research or during the initial part of the workshops, distrust regarding this type of methodology was noticed when some participants expressed feelings like: "I already know, you'll do your research and go away" or "I have already taken part many times here and no changes happened." CL proposal is different, and the discussion about its principles during the workshops was a strategic moment in improving research odds of reaching its objectives.

These presentation workshops can be useful tools for introducing this new approach, especially when an external player such as the one described here enforces changes on a company. This strategy contributed to overcome participants' mistrust. From a participation determined by the company, they evolved to important moments of agency and brought up to discussion contradictions within the activity system, as well as contradictions with other systems with which they interact or act simultaneously. Through the adopted dynamics, participants managed to view the network of relationships among different sectors and departments – systems of activities that have to work together.

Considering that other improvements and changes had been attempted in the company before, participants of the workshops recommended to researchers to seek top hierarchy's recognition and endorsement for the workers to take part in this research. That's why they suggested the meeting with their president – to foster means for an actual implementation of future solutions to be developed during CL sessions.

The organizational silence noted during ethnographic data collection was overcome through the workshops on MAPA and CL. We believe that they marked a turning point and were an important strategy in creating a climate of trust and bonding, and also in engaging company's employees into the research.

This study was financed by the Brazilian Labor Prosecution Office (MPT).

References

Almeida, I. M., & Vilela, R. A. G. (2010). *Modelo de análise e prevenção de acidente de trabalho – MAPA*. Piracicaba: CEREST.

Almeida, I. M., Vilela, R. A. G., Silva, A. J. N., & Beltran, S. L. (2014). Modelo de Análise e Prevenção de Acidentes - MAPA: ferramenta para a vigilância em Saúde do trabalhador. *Ciência & Saúde Coletiva, 19*(12), 4679–4688. https://doi.org/10.1590/1413-812320141912.12982014.

Engeström, Y. (1987). *Learning by expanding: An activity-theoretical approach to developmental research*. Helsinki: Orienta-Konsultit.

Engeström, Y. (2007). In H. Daniels, M. Cole, & J. M. Wertsch (Eds.), *Putting Vygotsky to work: The change laboratory as an application of double stimulation* (pp. 363–425). Cambridge: Cambridge University Press.

Engeström, Y. (2015). *Learning by expanding: An activity-theoretical approach to developmental research*. New York: Cambridge Press.

Engeström, Y., Virkkunen, J., Helle, M., Pihlaja, J., & Poikela, R. (1996). The change laboratory as a tool for transforming work. *Lifelong Learning in Europe, 1*(2), 10–17.

Ferreira, L. L. (1993). Análise Coletiva do Trabalho. *Revista Brasileira de Saúde Ocupacional, 78*(21), 7–19.

Jackson Filho, J. M., Vilela, R. A. G., Garcia, E. G., & Almeida, I. M. (2013). Sobre a "aceitabilidade social" dos acidentes do trabalho e o inaceitável conceito de ato inseguro. *Revista Brasileira de Saúde Ocupacional, 38*(127), 6–8. https://doi.org/10.1590/S0303-76572013000100001.

Lopes, M. G. R., Vilela, R. A. G., & Querol, M. A. P. (2018). Anomalies and contradictions in an airport construction project: A historical analysis based on Cultural-Historical Activity Theory. *Cadernos de Saúde Pública, 34*, ePub 00130816. https://doi.org/10.1590/0102-311X00130816.

Nuutinen, M., & Norros, L. (2009). Core task analysis in accident investigation: Analysis of maritime accidents in piloting situations. *Cognition, Technology & Work, 11*(2), 129–150. https://doi.org/10.1007/s10111-007-0104-x.

Osório, C., Machado, J. M. H., & Minayo-Gomez, C. (2005). Proposição de um método de análise coletiva dos acidentes de trabalho no hospital. *Cadernos de Saúde Pública, 21*(2), 517–524. https://doi.org/10.1590/S0102-311X2005000200018.

Querol, M. A. P., Jackson Filho, J. M., & Cassandre, M. P. (2011). Change Laboratory: uma proposta metodológica para pesquisa e desenvolvimento da aprendizagem organizacional. *Administração: ensino e pesquisa., 11*(4), 609–640. https://doi.org/10.13058/raep.2011.v12n4.143.

Virkkunen, J., & Newnham, D. (2013). *The Change Laboratory: A tool for collaborative development of work and education*. Rotterdam: Sense Publishers.

Yoon, Y. S., Ham, D. H., & Yoon, W. C. (2016). Application of activity theory to analysis of human-related accidents: Method and case studies. *Reliability Engineering & System Safety, 150*, 22–34. https://doi.org/10.1016/j.ress.2016.01.013.

Chapter 5
One Activity, Two Objects: Preparatory Phase of Change Laboratory at an Assistance Center for Adolescents in Conflict with the Law

Luciana Pena Morgado, Rodolfo Andrade de Gouveia Vilela, Mara Alice Conti Takahashi, Marco Antonio Pereira Querol, Silvio Beltramelli Neto, and Sandra Regina Cavalcante

Abstract The present chapter describes the preparatory phase of a Change Laboratory formative intervention targeting CASA Foundation's activity. This is a public institution charged of socio-educational measures for adolescents in conflict with the law. Our study was triggered by a demand from the Labor Prosecution Office following complaints of CASA employees exposed to occupational physical and mental hazards. Initial ethnographic research led to a preliminary hypothesis, namely, that CASA Foundation is undergoing a complex shift from a containment-based activity system to a socio-educational activity model. This shift is attended by contradictions which hinder the full implementation of the new model. In this chapter we describe the results of the phase of negotiation and ethnographic data collection, which is the first step before application of the CL method targeting the institutional cycle of expansive learning.

L. P. Morgado (✉) · R. A. G. Vilela · M. A. C. Takahashi · S. R. Cavalcante
Department of Environmental Health, School of Public Health, University of São Paulo, São Paulo, SP, Brazil

M. A. P. Querol
Department of Agronomic Engineering, Federal University of Sergipe, São Cristóvão, SE, Brazil

S. B. Neto
Labor Prosecution Office and Catholic Pontifical University, Campinas, SP, Brazil

R. A. G. Vilela et al. (eds.), *Collaborative Development for the Prevention of Occupational Accidents and Diseases*, https://doi.org/10.1007/978-3-030-24420-0_5

5.1 Introduction

In Brazil, health promotion actions in the workplace are usually designed from the perspective of managers, who as a rule do not seek the employees' views on their actual working conditions (Maeno and Paparelli 2013; Cardoso and Morgado 2015). In ergonomics, work activity is understood as the mobilization of the physical, cognitive, emotional, and social abilities of each worker to meet the employer's demands (Daniellou 2005).

When the determinants of the health–disease process are not duly identified, actions are restricted to correct effects but do not actually reduce the rates of illness and work-related accidents. Empowering workers for active participation is essential to change this situation (Clot 2010).

What might one say, then, of complex total institutions, namely, places of work and residence where a great number of similarly situated people, cut off from the wider community for a considerable time, together lead an enclosed, formally administered round of life (Goffman 1992)? Are inmates a socially excluded population? Do the institutions' employees use instruments with repressive intent naturally associated with the activity object?

In total institutions such as Adolescent Socio-educational Assistance Center Foundation (CASA),[1] there is a fundamental division between the large population subjected to control, i.e., the inmates,[2] and the small supervision staff. Each group tends to see the other through narrow and hostile stereotypes: the management sees inmates as bitter, reserved, and unreliable, while the latter see the former as intransigent, arbitrary, and petty (Goffman 1992).

However, the relational dynamics within this setting does not unfold exclusively between these two poles. As a function of the long period of time they live together, with the consequent exchanges and bonds, inmates organize into several groups with rules of their own, which makes life in total institutions even more complex (Diuana et al. 2008). To this we should further add the work organization model currently in force based on managerial mechanisms for the control of individuals. Employees are conditioned to perform their job just to accomplish goals and earn bonuses, which interferes with the achievement of the expected outcome, i.e., the social reintegration of adolescents (Metzger et al. 2012; Takahashi et al. 2015).

A survey and discussion of the various narratives and representations on health within a space intended for confinement and encompassing the various social roles (inmates, employees, management, and community) are relevant, as it allows denaturalizing representations and arbitrary practices and opens new possibilities of agency (Diuana et al. 2008).

The disparity between actors pointed out by Goffman (1992) and Diuana et al. (2008) represents a part of the conflict at CASA assistance centers. A position of privilege might cause estrangement among the various groups within the institution,

[1] In Portuguese: Fundação Centro de Atendimento ao Adolescente—CASA.

[2] In the present case, adolescents in conflict with the law.

resulting in violent situations involving both employees and inmates. Still according to these authors, the main concern is with the risk to personal integrity, which is strongly influenced by the formal and informal relationships which regulate life at the institution.

CASA Foundation has been the subject of academic studies conducted with different approaches usually within the context of social science, psychology, education, and health. A part of such studies explored the experience of coexistence at the institution from the adolescents' perspective (Almeida 2010; Moreira 2011; Vincentim 2005). Other studies focused on the functioning of the public school at the institution and hindrances to and contradictions in the attempts to apply socio-educational measures within a prison environment (Lopes 2006; Meireles and Zamora 2017; Yokomiso 2007).

One further group of studies performed analyses, some of them longitudinal, of changes in the child and adolescent protection legislation and their reflection on institutional modifications over time (Alvarez 1989; Passetti 2004; Lima 2010). Two academic studies conducted from this perspective—with 20-year interval between them—reported experiences of attempts at introducing changes and elucidated hindrances and challenges to structural transformations within the institution (Bierrenbach 1987; Lima 2006).

We also located studies which addressed the employees' social representations based on the practices, discourses, and instruments available in their work routine for application of socio-educational measures (Farias and Narciso 2005; Franco 2008; Lima 2014; Tavares et al. 2008). These studies explored facets of the object which allowed elucidating some obstacles and hints of contradictions which make actual application of the socio-educational measures indicated by the institution unviable.

The objectives of the present chapter are (1) to fill the gap in the knowledge about experiences of formative interventions as a part of a historical analysis of the development of the target activity as a public policy and (2) to describe the process of social construction of the demand and negotiation which granted us access to the institution to collect data on the daily functioning of an assistance center.

Collecting ethnographic data was important to support the pre-diagnostic hypothesis of tension in the activity system, which will be presented to the employees at a later time to establish together an effective diagnosis and develop the various phases of the cycle of expansive learning included in the Change Laboratory (CL) method (Engeström et al. 1996; Virkkunen and Newnham 2013).

5.2 Social Construction of the Demand

CASA Foundation is responsible for managing about 145 assistance centers and comprises about 10,000 employees charged of implementing socio-educational measures for about 9000 inmates aged 12–20 years and 11 months old in the state of São Paulo, Brazil (Nuprie 2018). The staff at the assistance center where the present study was conducted includes about 60 employees allocated to the following areas: director's office, technical coordination, administration, pedagogical staff, psychosocial staff, socio-educational support agents, cleaning, kitchen, laundry, and property security. About 60 adolescents are detained at this center.

Employees—except for the ones allocated to the administration, cleaning, kitchen, laundry, and property security—are charged of ensuring the physical and mental integrity of inmates along the full detention period, as well as of socio-educational actions aiming at their social reintegration. While the meaning of the term "social educator" cannot be found in the institutional manuals, the institution asserts that its mission is "to educate together with society for society; to educate for life" (Supsed 2014, p. 9). Other socio-educational activities, vocational training, music and visual arts, literature, and cooking courses are conducted in partnership with municipal, state, and nongovernmental organizations (NGOs) as part of their public policies.

Side by side with the socio-educational actions, containment instruments are applied to facilitate work. Whenever inmates move from one area to another, they are subjected to thorough inspection—about five times per day, on average—during which they are undressed and carefully examined by socio-educational support agents. Inmates must receive permission to go to the toilet or any other place and must circulate with the hands clasped behind the back. The aim of these practices is to discipline the body within a physical space intended for containment and a daily routine which does not promote positive agency between inmates and employees—nor even among the employees themselves, which thus favors the occurrence of illness among workers without proper resources to deal with such complex object. The demand which gave rise to the present study is a reflection of a hybrid institutional system comprising socio-educational and containment-centered practices which put the activity system under tension (see Figs. 5.1 and 5.2).

Fig. 5.1 Adolescents going to the bedroom. (Source: Marcos Santos/USP imagens)

Fig. 5.2 Image of adolescents' living space during the day to carry out school activities, art, and culture. (Source: Marcos Santos/USP imagens)

Many complaints of poor working conditions at CASA Foundation assistance centers were raised over time. A class action filed in 2003 by the Labor Prosecution Office (MPT) at labor courts enabled inspections of some assistance centers; workplace health and safety problems were indeed detected. Employees were found to be continuously exposed to infectious diseases due to lack of protective equipment, work accidents resulting from clashes with inmates, and institutional moral harassment. In addition, the sick leave rates were quite high.

In 2015 the employees of a CASA center filed a new complaint at MPT, in which they reported instances of physical and psychological violence, demanded better health and safety conditions, and claimed danger pay—conflict with inmates being considered hazardous working conditions.

In the prosecutor's view as per the previous complaints, the conventional and legal approaches to health and safety at work had not succeeded in explaining the complex interactions inherent to the target activity. As a result, such attempts had merely led to narrow-scoped corrective measures, while the relational model had remained unchanged. His hypothesis was that the containment-centered model adopted at CASA Foundation was a source of violence among the involved actors (inmates and employees): "both parties are caught in institutional violence." Given the complexity of the situation, the prosecutor chose to combine MPT methods with other systemic and formative approaches, such as CL, in the hope of achieving more consistent and long-lasting transformations. In other words, a viable path to transform the activity of reintegrating adolescents in conflict with the law to society.

Change Laboratory is a method which helps the involved actors develop new work models with the mediation of investigators. It is based on the principles of multi-voicedness, historicity, and systemic and dialectic vision, which help actors approach problems from the perspective of relations and interactions, rather than as fragments. It seeks to identify the systemic and historic origins of conflicts in a way the voice of actors is heard, and conflicting issues to be overcome are pointed out. It further enables retrieving successful experiences lost in time, which might be re-experienced, enriched, and consolidated as stable practices.

For this purpose, the School of Public Health, University of São Paulo, and the 15th Region of MPT signed a technical cooperation agreement to perform CL with CASA Foundation employees and inmates to formulate solutions together (São Paulo 2013). Due to several inmates' complaints, also a representative of the MPT, the Children Prosecution Office (MPE) was called to participate in this integrated interinstitutional action.

5.3 Negotiation for Workplace Access

The first step involved two important negotiations: One with MPE, MPT/child and juvenile court, which is responsible for monitoring the adolescents detained at the institution and the other involved the regional board and the CASA Foundation President's Office, to admit or not admit existence of the problem, present the CL method, and obtain authorization to perform data collection and apply the method at the assistance center mentioned in the 2015 complaint.

Double stimulation[3]was the strategy selected to help the participants recognize the problem. The first stimulus was a question posed to the managers: "What are CASA Foundation's main strengths and difficulties in the accomplishment of its mission in compliance with the institutional vision and values?" Following a discussion of the challenges which ought to be overcome to achieve the expected results, the second stimulus was applied, to wit, the suggestion to apply to CL as tool for formative development and participatory transformation of the ongoing activity model.

Double stimulation—as negotiation phase strategy to trigger reflections on strengths and difficulties—enabled managers to name challenges and achievements. Many challenges were reported to the investigators, such as difficulty to standardize procedures, absenteeism, and other system development problems. CL as second stimulus awoke hope among the managers: "This is a chance for us to change the Foundations' future," as one of the participants observed.

Authorization was obtained via a consent form, and it was agreed that data collection would be monitored all along the process by the involved institutions. The data would therefore provide the information needed by MPT, and at the same time

[3]Cognitive strategy used as a Change Laboratory resource to mediate between investigators and participants.

ground the process of participatory transformation as mirror data for activity system remodeling in the future CL sessions.

Meetings of investigators and employees of the target assistance center were scheduled to present the method and formulate the demand. Recognition of problems by the involved parties is an essential requirement for application of the CL method. It demands agreement between managers and volunteers interested in experiments at the local level. The employees listed problems in the execution of the activity, such as work overload, absenteeism, inordinate goals, punishment, and control. Then the phase of data collection began.

5.4 Data Collection Methods

Fieldwork took place from August 2016 through July 2017 in weekly visits to the target center/unit of analysis to collect empirical data; consult historical documents, legislation, articles, and films; and review institutional documents, such as political–pedagogical plans, materials developed by the strategic information production unit, and safety and discipline manuals.

During visits we also observed the daily routine at the institution, with focus on the interactions between inmates and employees. We performed semi-structured interviews with administration, social work, education, psychology, and health and safety professionals. Four focus groups were conducted with inmates. Formal interviews and focus group sessions were audio recorded and informal interviews and the investigators' impressions were entered on a field diary (see Table 5.1).

Individual and group interviews were performed at the center and lasted about 2 hours each. Due to structural limitations, the talks took place in offices, following authorization by the non-interviewed employees.

Bibliographic references were obtained from the Integrated System of Libraries of University of São Paulo (SIBI–USP), the dissertation database of Pontifical Catholic University (PUC–SP), and Scientific Electronic Library Online (SciELO). Other citations were identified in the selected studies. References were selected when analysis of their abstract or summary showed they were compatible with the subject of interest and addressed work and violence at resocialization facilities and their contradictions and challenges to the execution of socio-educational programs.

5.5 Results and Discussion of the Pre-diagnosis

Total institutions to confine adolescents in conflict with the law exclusively were established in Brazil starting in 1894. Their creation met a social demand triggered by the so-called Free Womb Law, which "freed" children of slaves after age 6 years old. Children who did not remain with their mothers as domestic workers at slave

Table 5.1 Summary of topics addressed in interviews

Function	Objective	Interviewee	Data collection method
Security coordinator	Professional and institutional trajectory; training to perform activity; activity strengths and difficulties; relationship with other areas; inclusion in reference team; expectations in regard to the study	One employee	Audio-recorded interview
Security coordinator	Procedures performed by the security coordinator	One employee	Recorded in field diary
Socio-educational support agent	Professional and institutional trajectory; training to perform activity; changes in activity over time; how they do their job; how they act in the reference group; activity strengths and difficulties; expectations in regard to the study	Two employees	Recorded interview
Socio-educational support agent	Professional and institutional trajectory; this informal interview took place in the laundry and included the participation of two inmates	One employee	Recorded in field diary
Socio-educational support agent	Professional and institutional trajectory; what led to file the complaint; training to perform activity; description of daily work; changes in activity over time; activity strengths and difficulties; expectations in regard to the study	Two employees	Audio-recorded interview
Center director	Professional and institutional trajectory; motivation to work at the center; how to run the center; implementation of procedures; relationship with employees and inmates; management methods	One employee	Audio-recorded interview
Social worker	Motivation to work at the institution; how each worker organize their work; how the articulation with the reference team is; whether they are able to meet the adolescents' expectations taken as reference; expectations	One employee	Audio-recorded interview
Administrative staff	Motivation to work at the institution; how they organize themselves at work; how they perceive bureaucratic procedures; expectations	Three employees	Audio-recorded interview
Pedagogical staff	Motivation to work in the institution; how each worker organize their work; how the articulation with the reference team is; whether they are able to meet the adolescents' expectations taken as reference; strengths and difficulties; expectations	Four employees	Audio-recorded interview

(continued)

Table 5.1 (continued)

Function	Objective	Interviewee	Data collection method
Healthcare staff	Professional trajectory; motivation to work at the institution; how they organize team work; how they articulate themselves with other professionals; how decisions are made together with the reference group; relationship with inmates; function strengths and difficulties; expectations	Two employees	Audio-recorded interview
Unit technical coordinator	Professional trajectory; motivation to work at the institution; relationship with employees and inmates; management methods to organize the reference team; expectations	One employee	Audio-recorded interview
Regional director	Information available; quantitative data requests; information management; if and how systematized information is transformed into prevention policies	One employee	Non-recorded interview
Regional pedagogical supervisor	How central guidelines are received; choice of courses; possibility to carry out autonomous activities; objectives of chosen courses	One employee	Recorded in field diary
Socio-educational support agent	Details on how they perform their activity	Seven employees	Focus group
Inmates	Trajectory at and outside the center; how they perceive their life at the center; conflict situations in the daily life at the center; expectations after release	Talks with 4 groups of 5–6 inmates at the cafeteria and classrooms	Recorded in field diary
Security superintendent	Professional trajectory; challenges to the implementation of measures; implementation of procedures; management possibilities and limits; what worked well and what does not work well in regard to management; socio-educational training for employees; articulation with other teams; activity strengths and difficulties	One employee	Audio-recorded interview
Education superintendent	Professional trajectory; challenges to the implementation of measures; implementation of procedures; management possibilities and limits; what worked well and what does not work well in regard to management; socio-educational training for workers; activity strengths and difficulties; what references might serve to ground guidelines; articulation with security; how suggestions are forwarded to units; partnerships; post-detention programs	One employee	Audio-recorded interview

owners' homes were sent to institutions designed for this socially excluded population. In time these institutions underwent discursive reformulation and developed instruments to meet the demand through less oppressive rules. In the present time, they advocate a socio-educational model focused on social reintegration. In the state of São Paulo, these institutions are currently administered by CASA Foundation. They underwent changes over time including paradigm shifts to address an object (adolescents in conflict with the law) under constant transformation.

The object of the earliest institutions of this type in the country were children and adolescents, many of whom had been abandoned. Their activity system was characterized by rules and a division of labor centered on containment, under a hygienist view, instruments being intended to prepare inmates for the production system, which demanded cheap and docile labor (Passetti 2004). Along this period, inmates were forced to perform strenuous agricultural tasks at reformatories at the expense of formal schooling, an approach that did not contribute to their full development. Illegal activities—involving a complex exchange network "through which stolen goods, bodies, drugs, and profit circulated"—were more effective than the efforts to include inmates in the labor market (Passetti 2004, p. 356).

New guidelines for the functioning of these institutions were formulated in 1964, when Brazil was under a dictatorship. According to the new nationwide policy, all guidelines were centralized, and a new, interdisciplinary and less repressive approach was formulated. However, it seems it was never fully implemented, as suggested by complaints raised to national and international human rights organizations by inmates, their relatives, and other actors involved in the protection of adolescent rights.

The analyzed documents show that the interdisciplinary approach failed, given the predominance of the containment instruments inherited from the older reformatories, which were reproduced in the model adopted for the Child Well-being State Foundation (FEBEM).Within such context, assistance was only sporadically provided when the security department granted the due authorization. "When the security staff judged that transferring an intern was risky, he would not receive any assistance on that day. One team's actions interfered with the other's"—observed by an interviewee (alluding to the security staff and the technical reference team).

A series of riots and criticism raised by the community led to change the institution's name and methods in 2005, when the Inter-American Court of Human Rights (IACHR) requested measures against the Brazilian government as a function of reports of torture and riots at the assistance centers (CIDH 2005).

After State Law no. 12,469/2006 was passed, assistance units for adolescents were renamed Socio-educational Assistance Center Foundation (CASA) to meet IACHR's recommendations and in compliance with the federal legislation (Child and Adolescent Statute—ECA—and the Federal Constitution). Priority was given to changes in instruments, rules, and the division of labor concerning the assistance provided to inmates. Socio-educational actions, compulsory schooling, and arts (graffiti, guitar playing, theater), culture (literature), and vocational training (customer service, waiting staff, telemarketing operator) courses were established, aiming at the social reintegration of inmates (Brasil 1988, 1990, 2012).

Thus the paradigm shift which led to ECA introduced considerable changes into the object, inasmuch as children and adolescents earned the status of subjects of rights. Changes were not only of name—from FEBEM to CASA—but had profound impact on the institution's structure. In 2006 the system was decentered into units at medium-sized municipalities, and the large complexes which had lodged about 2000 inmates were closed.

One further change targeted the division of labor. Starting 2014, the staff includes reference teams which assist inmates using interdisciplinary methods. The older FEBEM monitors became socio-educational support agents. The overall goal is to have all professionals work together in the application of socio-educational measures through individual and collective dialogue with inmates on their trajectory and experiences and staff meetings to discuss cases.

However, the new rule for division of labor did not suffice for the employees to integrate in an organic manner and their functions to be at a same hierarchical level—i.e., all the professionals being equally responsible for the reintegration of inmates. According to one interviewee, "We'd have a meeting on some kid, and if a [member of the] reference [staff] isn't there, it takes place anyway…then we refer it to such and such."

In this situation there are two objects in conflict: containment as additional punishment to deprivation of liberty and socio-educational measures aiming at social reintegration. Each object has mediation devices of their own, which behave as opposing forces: the use of collective punitive devices on the one hand and instruments such as formal education and vocational and cultural courses, dialogue, and interdisciplinary discussions of cases leading to individualized care on the other, as shown in the following statements:

> Each area has a vision of its own, they don't speak a same language, talks are needed among the staff [...] It shouldn't be like this, unfortunately. Some of the kids' provocations are subtle, they can really get you out of balance....You try and guide a kid, but he comes and says: "I don't care, I really do it…." You get mad and would like to slap him.
>
> It's a circuit that doesn't close... there's no interdisciplinarity whatsoever in the work done... each one pulls to their side...need to close it... we work in isolation... kids see us as having the power to punish....

The employees' reasons to seek jobs at CASA Foundation have very little to do with the institutions' aims. The following statements by socio-educational agents illustrate this point:

> Need. Some colleagues of mine worked at the Foundation. I sought a position to get job stability and everything else. I had already attempted other public service jobs, police officer, prison guard (Socio-educational agent 1).
>
> Everybody got in the Foundation due to pain. There were no jobs, so, let's go (Socio-educational agent 2).

Lack of proper training to deal with the daily complex situations makes the orientations provided by older "patio" (courtyard)[4] employees—trained in the former FEBEM model—prevail:

[4] Courtyard employees are the ones who spend more time with interns in living spaces for inmates.

> All the orientation given during training concerned security. They used the term "foot to the neck" a lot… training included a bit of terrorism: "When the kid jumps on, go for him."

Training for newly hired employees merely consists of 10 days of lectures on administrative issues and labor rights. Humanized approach to the containment of adolescents is not a priority, which compels employees to adopt practices out of line with the new CASA adolescents' profile:

> Teens used to be more tolerant in the past […] there were a lot of homeless children, they were sent to assistance homes or readmitted to the Foundation. Now, the kids confront us in any situation, and their families are complaining more and more to the MPT (Socio-educational agent 3).

When an activity is centered on the provision of care to others, interactions demand more involvement of the affective dimension of individuals (Vilela et al. 2010). Reintegration requires some development of bonds with inmates, but this goal is hindered by the many difficulties not addressed in the formative process. One example is provided by situations in which the relationships among inmates—who develop rules of their own to resist detention—reflect on their relationship with CASA employees.

When the available support does not meet the actual learning needs, the involved actors begin improvising, just to solve immediate problems, which results in much insecurity given the continuous exposure to conflict:

> Too often you can't handle the kids, it's exhausting. It's a prison-like place, the structure messes with everybody. Some employees can't take it and resign after a short while. It's a weird atmosphere. Very bad vibes (Social worker 1).

Then, this is a hierarchical organization centered on priorities which do not correspond to the actual needs, but administrative issues receive more attention than proper care to inmates:

> Too much red tape, too much happens at the same time... hard to administer restorative measures, for example. Socio-educational work is diluted in all that red tape (Social worker2).

Given that the study is in the step of initial data collection, our hypothesis is that there is tension within and between the elements of the institution's activity system (see Fig. 5.1). This tension gives rise to situations in which violence manifests in the relationship of employees with both inmates (subject and object) and management (subject and rules). Inclusion of productivity mechanisms in daily work—such as organizational harassment to accomplish inordinate goals, with consequent increase of the pace of work—contributes to the internal conflict, which in turn leads to work-related accidents and diseases.

The intended purpose of dialogue as a tool of mediation within the context of social education (subject, instrument, and object) is to promote the social reintegration of adolescents. However, it seems that only a part of the staff appropriated the

---

As a rule, these employees develop rules of their own to deal with inmates, as the formal guidelines do not cover the full complexity of the activity.

objective and without proper conditions to accomplish it. As a result, different perceptions and meanings develop for one and the same work (subject and division of labor). Several activity system elements are linked to older practices and thus behave as opposing forces in everyday work and create an impasse for workers—socio-educational agents and technical reference staff—derived from lack of interaction.

The new model seems to want to develop and consolidate, while the older model resists. While both employees and inmates complain of violence, repression, and little room for dialogue, they are caught by violence: they "are subjected to the very same rules, placed on opposite sides, and probably feeling much fear" (Tavares et al. 2008, p. 41) (see Fig. 5.3).

5.6 Final Considerations

All throughout its history, violence was a resource systematically used by the dominant classes to maintain the social organization in Brazil (Zaluar 1996; Minayo 2006). In total institutions such as CASA Foundation, the subject–object relationship is impregnated by several forms of violence: intimate inspection of inmates, additional punishment, verbal abuse in times of tension, inordinate goals, and organizational moral harassment of employees.

The shift from the FEBEM to the CASA Foundation model is marked by conflict having distal causes inherited from the reformatory period. At that time, the object had a different purpose formulated in response to the social demand posed by abandoned youths, either delinquent or not but seen as dangerous by society. The poor national and international repercussion of the resulting practices led to changes in the institutional rules. Such changes were, however, introduced top-down into the activity system, without the participation of the employees, and detached from the

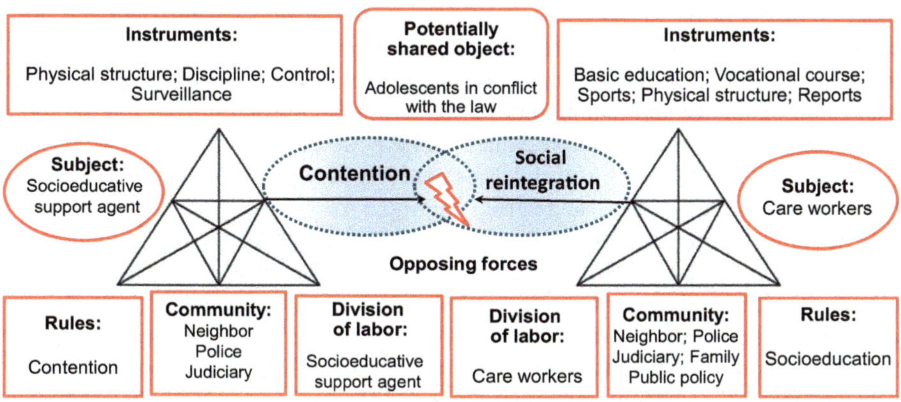

Fig. 5.3 Activity system—CASA Foundation: two objects for the one and the same activity. (Adapted from Engeström 2009)

evolutionary cycle of the object, which by this time began to demand new opportunities of social inclusion to avoid slipping into the world of illegality (Bierrenbach 1987; Sader 1987).

CASA employees have different perceptions of inmates. The socio-educational agents—who see themselves as mere security guards—are conditioned to systematically mistrust inmates and to restrain any behavior rated "suspicious." In turn, the role of the interdisciplinary staff is to apply the socio-educational guidelines into practice—which, as a fact, could never be fully implemented due to their divergences from the still predominant containment rules. Given the presence of two objects in dispute, two activity systems coexist in opposition within a same time and space and cause mutual hindrances to the accomplishment of the outcomes expected by the institution.

We thus believe there is a cycle of expansive learning in progress to meet the social demands of human and social rights affirmative action policies. The existing contradictions create security disturbances and conflict in relationships. On one side there is the resistance opposed by the older reformatory model, which involves application of additional punishment since it is assumed that this type of institution is intended to restrain a population rated unredeemable. On the other, there is a model which sees institutions of this type as educational facilities with pedagogical, assistance, and vocational strategies for the reintegration of adolescents made vulnerable by exclusionary economic models and deprivation of fundamental human rights.

The same contradictions which account for the disturbances in the systems also explain their development together with the attempts at improving the model. Historical development analysis allows visualizing a prospective trend, i.e., a zone of proximal development—a graphic representation which includes collective possibilities of future development through the formulation of new concepts and new ways of thinking on work practices leading to expansive learning of the object (see Fig. 5.2).

This model might yield the expected results, namely, relevant changes in all the other elements of the activity system: abandonment of the tacit physical and psychological violence rules in the relationship with inmates; a division of labor which links the various professional categories within a single shared pedagogic plan; expansion of family and community participation; and promotion of intersectoral work with other municipal and regional public policies, such as the ones focusing on health, education, housing, social assistance, employment, and income and regulatory agencies such as MPT and MPE (see Fig. 5.4).

The dialectic contradiction at the institution (social education vs. contention) became a hindrance to the accomplishment of the goals of the new public policy aiming at making adolescents protagonists of their own stories, which is a relevant and urgent social task. A participatory process designed with and targeting both workers and adolescents has potential to bring structural changes about to consolidate new internal and community relationships, thus unifying the activity system for positive agency between subject and object.

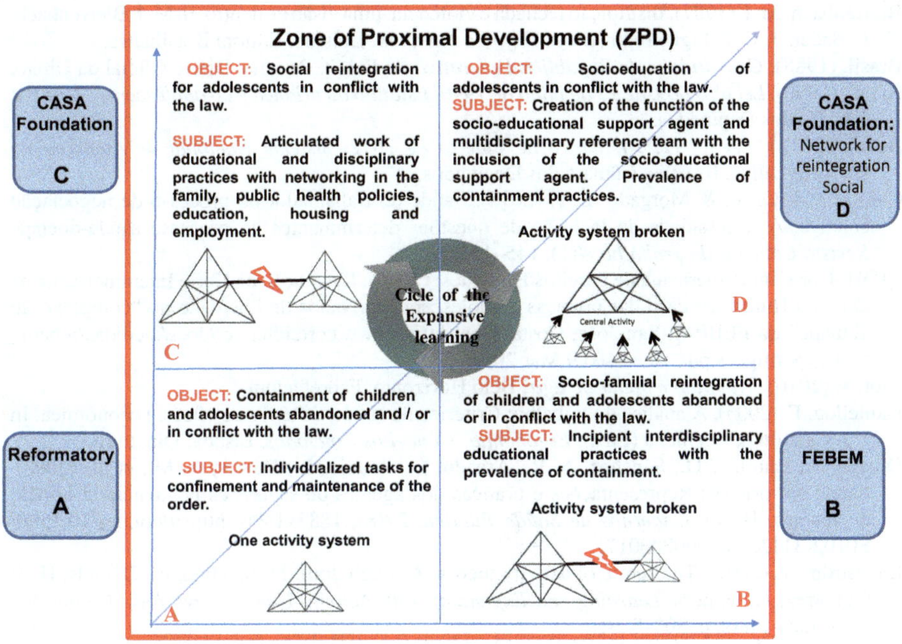

Fig. 5.4 Zone of proximal development (ZPD). (Adapted from Virkkunen and Newhman 2013)

The assumption underlying the testing of new models is to make work more meaningful and improve its effectiveness and the employees' and adolescents' health and safety conditions, based on the possibility to produce knowledge and changes at one time. The overall aim is for the CL training intervention to enable professionals to build a shared structure likely to overcome existing antagonisms and contribute to health promotion at the institution. The reason is that this method affords space for dialogue between parties which usually do not communicate with each other.

This study was financed by the Brazilian Labor Prosecution Office (MPT) and the Coordenação de Aperfeiçoamento de Pessoal de Nível Superior—Brasil (CAPES)—Finance Code 001.

References

Almeida, B. G. M. (2010). *A experiência da internação entre adolescentes: Práticas punitivas e rotinas institucionais [dissertation]*. São Paulo: Universidade de São Paulo. https://doi.org/10.11606/D.8.2010.tde-08022011-144629.

Alvarez, M. C. (1989). *A emergência do código de menores de 1927: Uma análise do discurso jurídico e institucional da assistência e proteção aos menores* [dissertation]. São Paulo: Universidade de São Paulo.

Bierrenbach, M. I. (1987). Instituição fechada e violência: uma visão de dentro. In M. I. Bierrenbach, E. Sader, & C. P. Figueiredo (Eds.), *Fogo no Pavilhão*. Brasília: Editora Brasiliense.
Brasil. (1988). *Constituição da República Federativa do Brasil*. Brasília: Diário Oficial da União.
Brasil. (1990). *Lei n° 8.069, de 13 de julho de 1990. Estatuto da criança e do adolescente*. Brasília: Câmara dos Deputados.
Brasil. (2012). *Lei n° 12.594, de 18 de janeiro de 2012. Sistema Nacional de Atendimento Socioeducativo*. Brasília: Câmara dos Deputados.
Cardoso, M. C. A., & Morgado, L. P. (2015). Saúde do trabalhador no processo de negociação coletiva: A necessidade de inclusão de questões determinantes do processo saúde-doença. *Revista Ciências do Trabalho, 4*(1), 145–156.
CIDH -Corte Interamericana de Direitos Humanos. (2005). Resolução da Corte Interamericana de Direitos Humanos: caso das crianças e adolescentes privados de liberdade no "complexo do Tatuapé" da FEBEM. San Jose, Costa Rica. http://www.corteidh.or.cr/docs/medidas/febem_se_01_portugues.pdf. Accessed 9 Mar 2017.
Clot, Y. (2010). *Trabalho e poder de agir*. Belo Horizonte: Fabrefactum.
Daniellou, F. (2005). A análise do trabalho: Critérios de saúde, critérios de eficácia econômica. In J. J. Castillo & J. Villena (Eds.), *Ergonomia: Conceitos e métodos*. Lisboa: Dinalivro.
Diuana, V., Lhuilier, D., Sánchez, A. R., Amado, G., Araújo, L., Duarte, A. M., et al. (2008). Saúde em prisões: Representações e práticas dos agentes de segurança penitenciária no Rio de Janeiro, Brasil. *Cadernos de Saúde Pública, 24*(8), 1887–1896. https://doi.org/10.1590/S0102-311X2008000800017.
Engeström, Y. (2009). The future of activity theory: A rough draft. In: Sannino, A., Daniels, H. & Gutierrez, K., editors. *Learning and Expanding with Activity Theory*. New York: Cambridge University Press. p. 303–328.
Engeström, Y., Virkkunen, J., Helle, M., Pihlaja, J., & Poikela, R. (1996). The Change Laboratory as a tool for transforming work. *Lifelong Learning in Europe, 1*(2), 10–17.
Farias, P. S., & Narciso, L. J. (2005). *Cadeia de chocolate: Os funcionários da FEBEM falam*. São Paulo: Arte andciência.
Franco, E. M. (2008). *Uma casa sem regras: Representações sociais da FEBEM entre seus trabalhadores* [these]. São Paulo: Universidade de São Paulo. https://doi.org/10.11606/T.47.2008.tde-06062008-170921.
Goffman, E. (1992). *Manicômios, prisões e conventos*. São Paulo: Perspectiva.
Lima, R. C. P. (2006). Mudanças das práticas sócio-educativas na FEBEM-SP: As representações sociais de funcionários. *Psicologia & Sociedade, 18*(1), 56–62. https://doi.org/10.1590/S0102-71822006000100008.
Lima, C. N. (2010). *O fim da era FEBEM: Novas perspectivas para o atendimento socioeducativo no Estado de São Paulo* [dissertation]. São Paulo: Universidade de São Paulo. https://doi.org/10.11606/D.48.2010.tde-09032010-151702
Lima, J. V. (2014). *Entre o recuperável e o estruturado: Classificações dos funcionários de medida socioeducativa de internação acerca do adolescente em conflito com a Lei* [dissertation]. São Paulo: Universidade de São Paulo. https://doi.org/10.11606/D.8.2014.tde-14012015-121821
Lopes, J. S. (2006). A escola na FEBEM – SP: Em busca do significado [dissertation]. São Paulo: Universidade de São Paulo. https://doi.org/10.11606/D.47.2006.tde-09042007-153821
Maeno, M., & Paparelli, R. (2013). O trabalho como ele é e a saúde mental do trabalhador. In M. A. Silveira (Ed.), *Inovação para o desenvolvimento de organizações sustentáveis: Trabalho, fatores psicossociais e ambiente saudável*. Campinas: Centro de Tecnologia da Informação.
Meireles, C. C., & Zamora, M. H. (2017). Sistema socioeducativo e tecnologias de poder: Análises preliminares sobre os agentes socioeducativos. *Revista Adolescência e Conflitualidade, 16*(1), 17–25. https://doi.org/10.17921/2176-5626.n16p17-25.
Metzger, J. L., Maugeri, S., & Benedetto-Meyer, M. (2012). Predomínio da gestão e violência simbólica. *Revista Brasileira de Saúde Ocupacional, 37*(126), 225–242. https://doi.org/10.1590/S0303-76572012000200005.
Minayo, M. C. S. (2006). *Violência e saúde*. Rio de Janeiro: Fiocruz.

Moreira, F. M. (2011). *Cadeias dominadas: Dinâmicas de uma instituição em trajetórias de jovens internos* [dissertation]. São Paulo: Universidade de São Paulo. https://doi.org/10.11606/D.8.2012.tde-15062012-153941
Nuprie. (2018). Boletim estatístico semanal: Posição 02/02/2018. São Paulo.
Passetti, E. (2004). Crianças carentes e políticas públicas. Criança e criminalidade no início do século. In M. D. Priore (Ed.), *História das crianças no Brasil* (4th ed.). São Paulo: Contexto.
Sader, E. (1987). Democracia é coisa de gente grande? In M. I. Bierrenbach, E. Sader, & C. P. Figueiredo (Eds.), *Fogo no pavilhão*. Brasília: Editora Brasiliense.
São Paulo. (2013, 13 Jun). Em defesa da saúde do trabalhador. *Diário Oficial Poder Executivo*. Sec 1:IV.
Supsed - Superintendência de Segurança e Disciplina. (2014). *Fundação CASA. Conceitos, diretrizes e procedimentos*. São Paulo.
Takahashi, M. A. B. C., Montanher, P. J. S., Silva, A. J. N., & Vilela, R. A. G. (2015). Destructive management behavior and organizational pathology. Reflections for an articulated intervention between academic groups, services and trade union movements. *Labour Sciences Journal, 4*(1), 27–47.
Tavares, E. P. N., Magalhães, F. B., Marques, P. H., Carmo, J. C. V., & Santos, T. L. F. (2008). *O trabalho dos monitores na Febem*. São Paulo: Fundacentro.
Vilela, R. A. G., Silva, R. C., & Jackson Filho, J. M. (2010). Poder de agir e sofrimento: Estudo de caso sobre agentes comunitários de saúde. *Revista Brasileira de Saúde Ocupacional, 35*(122), 289–302. https://doi.org/10.1590/S0303-76572010000200011.
Vincentim, M. C. G. (2005). *A vida em rebelião: Jovens em conflito com a lei*. São Paulo: Hucitec.
Virkkunen, J., & Newnham, D. S. (2013). *The Change Laboratory: A tool for collaborative development of work and education*. Rotterdam: Sense Publishers.
Yokomiso, C. T. (2007). *Violência e descontinuidade psíquica: Um estudo sobre a Fundação CASA* [dissertation]. São Paulo: Universidade de São Paulo. https://doi.org/10.11606/D.47.2007.tde-15012008-155933
Zaluar, A. (1996). *Da revolta ao crime S/A* (2nd ed.). São Paulo: Moderna.

Chapter 6
Conflict in a Formative Intervention at a Public School: Lessons for Researcher-Interventionists

Amanda Aparecida Silva-Macaia, Marco Antonio Pereira Querol, Frida Marina Fischer, and Rodolfo Andrade de Gouveia Vilela

Abstract The present chapter describes an intervention conducted at an elementary school in the São Paulo Metropolitan Area, Brazil. According to the occupational health and safety service and the Municipal Secretariat of Education, the demand for a Change Laboratory (CL) was triggered by disturbances in the school routine and teacher sickness absenteeism. These disturbances interfered with the students' learning. We also discuss conflicts and challenges which developed along the CL and eventually led to its discontinuation. We emphasize innovative actions implemented by the interventionists to solve the problems and continue the process of change. Final comments and suggestions address the researchers' learning process.

6.1 Brazilian Educational System and Context of Intervention

A common characteristic of educational systems since the 1990s in Latin America, including Brazil, is the development of policies to promote social justice based on principles of efficacy proper to the capitalist production model (Oliveira 2007). Different from simultaneous developments in Europe and North America in the same period, the quest for social justice in Latin America was based on expanding education—the actions initially implemented to democratize the access to education precisely targeted mass expansion.

A. A. Silva-Macaia (✉) · F. M. Fischer · R. A. G. Vilela
Department of Environmental Health, School of Public Health, University of São Paulo, São Paulo, SP, Brazil
e-mail:as.amanda@gmail.com

M. A. P. Querol
Department of Agronomic Engineering, Federal University of Sergipe, São Cristóvão, SE, Brazil

R. A. G. Vilela et al. (eds.), *Collaborative Development for the Prevention of Occupational Accidents and Diseases*, https://doi.org/10.1007/978-3-030-24420-0_6

This approach did, indeed, enable access to education to new segments of the population, however, with limited resources and deterioration of the professionals' working conditions and work processes. Furthermore, the educational policies just sought to train the workforce, ensure a minimum income, and provide social assistance (IPEA 2000; Oliveira 2007).

In Brazil, these factors led to a new system of school organization built on administrative, financial, and pedagogical decentralization. This system resulted in an educational regulation model which seeks to reconcile global demands and local responses, although the latter are often insufficient. The new model granted more flexibility and autonomy to teachers and school managers but also brought greater demands, responsibilities and, paradoxically, higher control over them (Oliveira 2007).

This framework has been characterized as intensification and deterioration of the educators' work, expressed as suffering and illness among teachers and other professionals (Oliveira 2007; Assunção and Oliveira 2009; Garcia and Anadon 2009). Among the many aspects of work precariousness, the ones considered relevant for workers' health, teachers in particular, include long working hours; having to teach to large classes and at different schools; overlapping of roles; frequent rescheduling due to high absenteeism rates; perceived lack of control over tasks; misalignment between pedagogic approaches and criteria to establish school classes; and lack of recognition by parents and guardians (Assunção and Oliveira 2009).

In this chapter, we describe a Change Laboratory (CL) (Engeström et al. 1996; Virkkunen and Newnham 2013) initially motivated by teacher absenteeism. We discuss problems in the activity—teaching to children aged 4–11 years old—with focus on the conflicts which emerged along formative intervention.

The school where CL was conducted is located in a suburb of São Paulo. From 2015 to 2016, it included 10 teachers, 3 managers, and about 840 students distributed across 3 shifts. There were 8 classes per shift, with about 35 students each. There was a teacher in charge of each class and an art and physical education teacher who taught all eight classes.

Teachers worked 25 h per week, 20 in the classroom and 5 preparing activities—two such hours mandatorily at the school. Students spent 4 h per day at school, always attended by some teacher. As a result, absenteeism had negative impact on the school routine and the students, i.e., the object of the teaching activity.

Our purpose is to elucidate the conflicts which arose along intervention and the challenges posed to the researchers-interventionists. What did they learn? To discuss the lessons they learned from this intervention, we had resource to audio files recorded in sessions and interviews, as well as to notes entered by the researchers on a field diary.

We begin by describing the challenges and conflicts which arose along the various CL phases. Next, we address the intervention as such, from the formulation of the problem which motivated it (absenteeism) to its ending. Next, we analyze several modes of collaboration to discuss challenges and conflicts related with the various activities and describe the strategies used to solve the problems which emerged along intervention. Finally, we summarize the lessons learned by the investigators and make recommendations for interventions targeting the analyzed activity in Brazil.

6.2 Making Conflicts Visible: From Negotiation to Testing New Ideas

Our intention here is not to discuss the conflicts which developed along intervention—which are the topic of other sections—but to describe the context within which they arose.

6.2.1 Development and Negotiation of the Demand

The analyzed CL resulted from an agreement between representatives of the Municipal Secretariat of Education of São Paulo and the third author. The main problem found was teacher absenteeism, which caused several disturbances in the school, such as need to reallocate students and hurry to find substitutes, with the consequent impairment of learning.

Discussions among the education system central managers eventually led them to ask why teachers were missing work. Some possible answers were the emotional overload inherent to the profession; violence at school or near areas; gender issues and the women's double burden as teachers/mothers; unavailability of substitute teachers; overcrowded classrooms; relational and learning difficulties of students; and loose legislation on tardiness and absenteeism. The managers further observed that absenteeism was higher among the younger teachers' days, possibly for having to care for their small children and having other jobs (Avellar 2017).

Teacher absenteeism triggered a demand for intervention when it became the subject of graduate research of an occupational health professional from the Secretariat of Education. Indeed, it was this professional who designed the CL-based intervention.

However, while the demand for intervention originated in both the central management and the investigators, the local managers of the target school did not participate in the negotiation of the intervention method. This fact would have substantial consequences, as we discuss later on.

6.2.2 Preparing Intervention and Collecting Ethnographical Data

The Secretariat of Education selected the school where intervention was to be performed based on the two criteria: smoother relationship with the school principal and better organization of the documents needed for analysis.

However, the local managers—principal, deputy principal, and pedagogical coordinator—did not understand the need for intervention, although they acknowledged absenteeism was a problem. According to them, the main disturbances were due to the need to reallocate students whenever a teacher was absent and the actions required to solve this and prevent other problems. One of such actions was to find substitute teachers.

Conflict/Challenge 1
The central management acknowledged the need for intervention, but local management did not. They felt compelled to participate in the intervention, which resulted in resistance to the perspectives of change.

The initial planning of sessions followed the objectives and deadlines of the graduate student's research project. Previous data collection included interviews, meetings, and documents on the school routine and absenteeism (see Table 6.1).

Data collection evidenced divergences in the ideas on absenteeism held by the school local managers and the teachers. The former saw it as a disturbance with negative impact on the students' education, while the latter considered missing work a labor right.

Conflict/Challenge 2
Doctoral dissertation behind schedule significantly limited data collection.

Based on the understanding of some of the problems, the following actors were invited to participate in CL sessions: the school principal, deputy principal, pedagogical coordinator, afternoon shift teachers, two safety technicians, and one Secretariat of Education representative. To preserve the channel of communication with the central management, we scheduled meetings every 2 months to provide feedback and discuss the intervention.

6.2.3 *Mapping the Situation: What Is the Problem?*

Three main problems were discussed in different CL phases, which reinforce the idea that although some particular types of expansive learning actions predominate in each phase of the cycle, there is a nonlinear connection of actions between phases, for which reason actions might reappear in other phases (Virkkunen and Newnham 2013).

In the first sessions, demand "absenteeism" as main problem was redefined as "teachers' exhaustion". Yet the problem was put into question once again to be reformulated as "relationship between teachers and the local management", which thus revealed a relevant side of exhaustion, especially among the teachers. At the beginning of the process, we asked about ongoing practices and problems at the

Table 6.1 Methodological procedures applied in the CL intervention at the school

	Procedures	Objectives	What and with whom?	
Phenomenological phase	Meetings	To establish contact to present the demand	3 meetings at the Municipal Secretariat of Education (1 meeting included the school principal)	
	Visits	To become familiarized with the school and the local management staff	1 visit which included the local management staff (principal, deputy principal, and pedagogical coordinator)	
	Individual interviews	To collect ethnographical data and formulate hypotheses on the target activity, main problems, historical contradictions, and innovations	Secretary of Education, 1 senior teacher, 1 junior teacher, 1 intern, 1 home babysitter of some students, and 1 mother	
	Document analysis		Documents on school routine and absenteeism	
Intervention phase	Sessions	To implement expansive learning actions	From August 2015 through June 2016, 28 1-h sessions	Teachers, local management (absent in some sessions), representatives of the OHS[a] and HR[b] departments
			August and September 2016, 2 2-h sessions	Local management
	Observations and interviews	To collect ethnographical data on the activities of local managers and teachers in the classroom and information on medical care in association with sick leaves	2 interviews with 2 teachers in the morning shift. Audio and video recording of the local managers' work routine and interview on the observed and recorded actions Observation of classes and record on field diary Interviews with occupational physicians	
	Meetings at the Secretariat of Education	To follow up intervention and collectively analyze some of its phases	5 meetings with the central management and OSH department (1 meeting included the school principal)	

[a]Occupational health and safety
[b]Human resources

school, which led the participants to describe a cycle of problems. Within this context, we further asked whether absenteeism was the core problem on which they should focus their attention to plan changes in the work routine.

To the local managers, absenteeism did not only cause several disturbances in the work routine but was a disturbance in itself. In turn, the teachers considered that missing work was a right justified by the working conditions. According to the managers, the cause of absenteeism was the teachers' lack of responsibility and commitment. Such view is common within the Brazilian culture and makes the vertical division of labor even more explicit.

In activity ergonomics, disturbances are the result of a structural tension between "prescribed work" and "actual work". Such discrepancy is due to the roles proper to actors in different hierarchical positions, even when their actions are centered on one and the same object.

Conflict/Challenge 3
Ensuring dialogue between two worlds, the activity of local managers and the activity of teachers, which although different were related.

6.2.4 *Analysis of Exhaustion and the Different Points of View of Teachers and Local Managers*

To identify the origins of the teachers' exhaustion, the participants were stimulated to analyze three aspects: the conditions under which exhaustion developed, its insertion into the activity system (AS), and its relationship with the development of the object. A timeline served as second stimulus in the sessions devoted to historical analysis (see Fig. 6.1). It helped the participants clarify some of the common

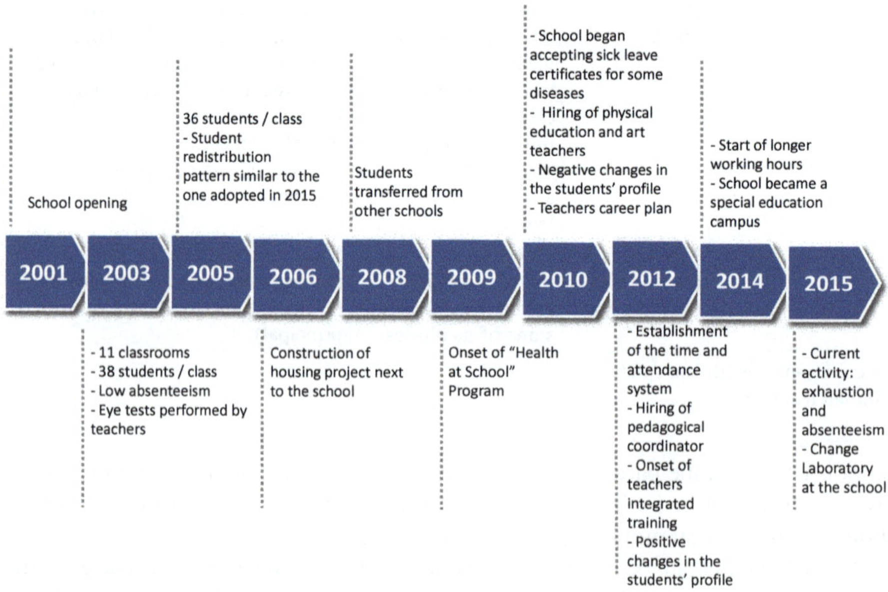

Fig. 6.1 Timeline formulated by the participants in the phase of historical analysis

justifications for exhaustion, which however did not hold when institutional changes were pointed. One resulting insight was that neither the large number of students in each class nor the way to solve the problems posed by teachers' absences alone were each one alone the reason of the increase in absenteeism and exhaustion, as they had not changed over time.

Among the changes which had taken place in the school, one was the arrival of new students in 2008 and 2010, with consequent demand of interactions mediated by new instruments, such as the so-called pedagogical mediation, namely, interventions targeting the students' attendance and learning. For instance, the local managers provided students orientation on homework and guided parents on how to deal with their children in regard to some problems, learning results, and school attendance. The new relationship between the local management and the community led to conflict with the teachers, given that both groups had their focus on the students' learning and the relationship with parents.

This mediation eventually resulted in conflits between the pedagogical orientations given by managers and those given by teachers, since they both didn't understand each other. In addition, since the parents were frequently called to the school or visited it spontaneously to make requests or objections, such mediation in time became a routine task.

The arrival of students with different needs and demands, the new learning instruments, and the new division of labor could have been the cause of the teachers' exhaustion. An excerpt from the records of the session devoted to historical analysis points to a change in the object:

> Coordinator (reading the record of the previous session): Teacher X says that then it was less exhausting, but now there are ... more projects being developed. You can feel a difference in the students' behavior by comparison to 2013. The younger parents educate differently, but the principal comes and says that then it was worse than now. Teacher Y says it's the other way around.
>
> Principal: About what?
>
> Coordinator: Time.
>
> (...)
>
> Principal: I said you used to have more assignments then than you have now.

Historical analysis enabled participants and interventionists to hypothesize contradictions. One was a primary contradiction within instruments, namely, the school daily reports and weekly plan. More specifically, the contradiction involved the number of forms to be filled and the quality of the information entered or even the very need to fill forms at all, given the short time available to prepare classes.

We stimulated the local managers and teachers to describe the ongoing notions and rules for daily reports and weekly plans. This strategy put both groups in confrontation and helped clarify some points of view.

Grounded on the understanding of contradictions, actual empirical analysis evidenced divergences in the work process of each group, as well as in their intentions in regard to the aforementioned instruments. The evidence gathered concerned the teachers' exhaustion and indirectly also their relationship with the local managers.

According to the teachers, the weekly plan could be modified or even abolished. The transcripts below illustrate the divergences between managers and teachers:

> Teacher X: (the weekly plan) (...) is the plan you prepare for classes, and is the same you record on the daily report form (…) only that you can't take it home. So I keep another, a third, at home, [I fill it] on weekends. And Mondays I copy [the data] to the weekly plan, and to the daily report form (…)
>
> Teacher A: Let me be honest, I haven't make it for quite a long time. I make [it] on my little notebook, and then copy [the data] to the daily report form (…) I believe the weekly plan could be something just for us, the teachers, not to be turned in, right? (…)
>
> Teacher X: (...) for official purposes, the daily reports would suffice, in regard to the management, the students, for you to bear in mind what you've already done, what you're going to do (…)
>
> Principal: the daily and the weekly plan, one complements the other.
>
> Coordinator: What I understood is what they'd like to throw away is the weekly plan that comes to me. What's the weekly plan? Your classes' planning, which I believe is extremely important, this is out of question, I believe it is (…) I agree with the principal in that one complements the other (…)

To the teachers, the daily report and weekly plan meant double work and exhaustion. However, according to the managers, they were useful to organize the work process and make it clear to the students. Although also other disturbances we could observe denoted several contradictions in the activity system, we call the attention to the participants' diverging views on the daily report and weekly plan, because they point to differences in the aspects of one and the same object. Some innovations were tested in regard to these instruments when the new school year began, but also new problems between the activities arose.

6.2.5 Conflicts Emerging During the Testing of New Ideas

The sessions devoted to suggest innovations were held at the end of the school year. We observed that specific changes in the usual activity could be tested immediately after the end of the holidays to motivate more systemic solutions. However, neither teachers nor local managers agreed on the need for changes and their control.

The central management had been long planning a new procedure for hiring substitute teachers. Therefore, it was decided that changes relative to the reallocation of students and work routine would be analyzed at the beginning of the new school year. However, by that time, the management had introduced changes in the daily report and weekly plan which the teachers had not liked at all. The managers' idea was to solve some of the problems addressed in previous sessions, which however they did without modeling solutions together with the CL participants. This situation evidenced problems in the relationship between the participating teachers and the school management. Indeed, the new forms suggested were just the tip of the iceberg.

The teachers observed that exhaustion was directly related to the work done jointly with the management and asked for the following CL sessions to be conducted separately.

6.3 Different Scripts for One and the Same Object: Grounds to Understand Challenges and Conflicts

Above we pointed out discursive divergences between teachers and managers in regard to instruments “daily report” and “weekly plan.” We analyzed these divergences based on the application of the notion of modes of collaboration (Raeithel 1983; Fichtner 1984 as cited in Engeström 2008) to the work process.

One of the ways of collaboration is coordiantion, when each people focuses on his or her own object (which might not necessarily be the same as the corresponding activity). One might even say on “their own interests” or “agenda.” In coordination, the script defines the time and place each one may speak. When cooperating, a problem draws their attention and actions on one and the same object; however, this does not mean they agree on a shared object. When communicating, actors have some idea of a shared object and consciously regulate the collaboration script together to adjust it to the object requirements.

Along the intervention period, both teachers and managers had to fill forms as a part of their work routine. Filling a large number of forms is a striking aspect of work in schools, as it is a setting characterized by a trend to record everything that occurs. This school had been favorably rated by the Secretariat of Education in regard to the organization of records, which suggests that the pressure to control and quantify results came from higher hierarchical levels. However, the teachers considered that filling so many forms competed with the teaching-learning actions. The management’s script defined good teaching based on the number of students, their success, school attendance, and teaching hours, i.e., all which has to do with the right to education. In turn, in the teachers’ script, good teaching depends of whether students grasp or not the intended content, which is not easy to measure in quantitative terms.

Figure 6.2 depicts two different scripts for teaching actions (attendance + pedagogical mediation and knowledge assimilation and reproduction) although the focus of both groups was on the education of students (object). The lack of coordination between scripts (represented by the X in the figure) was due to tension within the teaching actions (represented by the lightning bolt in the figure).

The teachers’ script (2) stated they should fill the daily report at the end of the work day on the premises, including information on the students’ attendance and the topics taught. The form should be filled with a pen, and erasing is not allowed. The topics should be identified by means of a code and the corresponding description in compliance with the parameters established by the Secretariat of Education. The result is standardized reports. Indeed, upon observing three teachers, we found similarities in their routine:

- The teachers interacted directly with students all throughout classes.
- They needed more than an hour to fill the daily plan, being constantly interrupted.

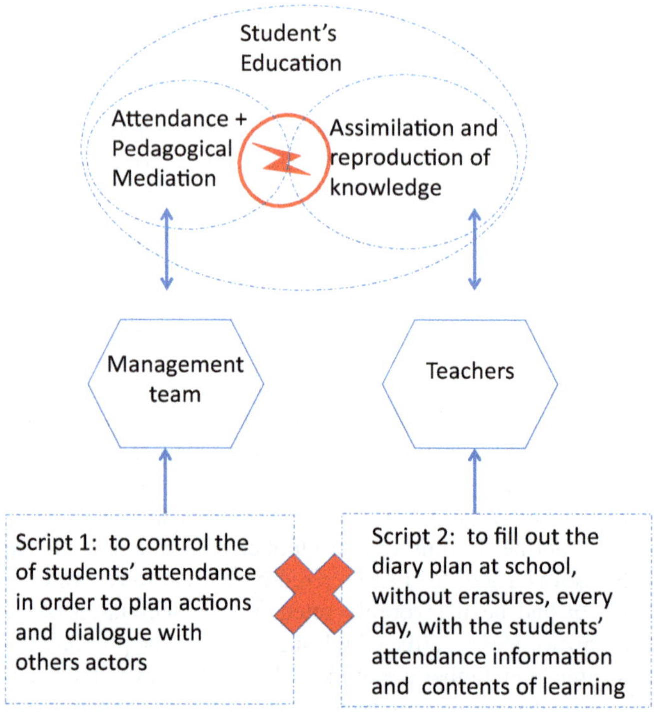

Fig. 6.2 Lack of coordination between different scripts and a shared object. (Adapted from Engeström 2008)

- They developed parallel activities and chose alternately not to record the attendance data or topics taught.

  One researcher made the following note:

  > [The teacher] took profit of the fact the students were responding questions [task] to fill the daily plan using the annual plan as basis. He observed that in the past he was requested to write less, but now he wasted more time, as he had to record more data; he had to dispel students' doubts and write long explanations [on the report form] at the same time.

According to the management's script (1), the information recorded by the teachers was used on several occasions, such as:

- Parents or relatives' visits to the school with complaints or questions or to request information on their children
- Request of information by another school about the attendance of some particular students
- Students´ attendance to plan joint actions with families
- Topics taught to plan actions with the teaching staff

The main reason for the conflict triggered by intervention was the managers' use of control instruments in opposition to the participatory and formative learning

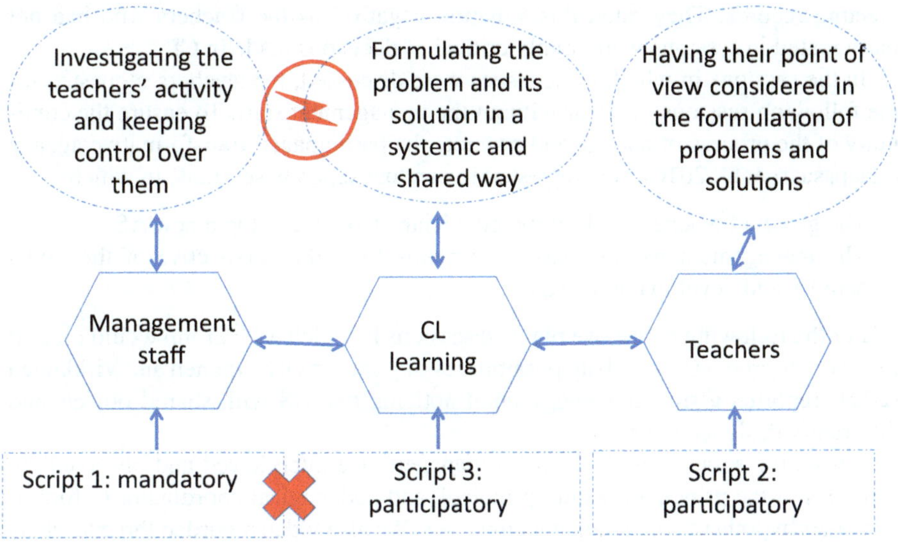

Fig. 6.3 Contradictions between scripts and objects of the activities considered in the intervention. (Adapted from Engeström 2008)

activity. The conflict between investigators and school managers first became clearly evident when the latter introduced innovations without discussing them with the teachers.

This and other aspects related to intervention revealed resistance to expansive learning in the shape of "sabotage" to CL. We may now describe the conflict as resistance derived from the contradiction between expected results (objects) and scripts for the management, expansive learning, and teachers' activities (see Fig. 6.3). We categorized the scripts as participatory or imperative as a function of how actors were to attain their results.

We conclude that the conflict developed between the management's expectations in regard to intervention and the actual intention of researchers upon suggesting application of CL. The participatory nature of the teaching activity ran against the imperative nature of some management activities. Researchers-interventionists and participants reshaped their initial expectations in regard to intervention once they realized they were restricted to obtain ready-made solutions from the researchers.

6.4 Learning from Intervention: Actions to Solve Conflicts

The teachers' discontent with the changes introduced by the managers appeared together with the development of mistrust between both groups. The teachers complained that everything said in sessions became available to all the teachers through meeting records. They rated this situation negative, as the teachers who had not participated in sessions were highly critical of decisions made in CL.

In the sessions in which these issues were discussed, the teachers requested for the following sessions to occur without the management staff. To ensure the continuity of the process of change and promote the participants' transformative agency (Haapasaari et al. 2016), we suggested to conduct separate sessions, in which:

- The group of teachers and safety technicians would develop a new AS.
- The management would analyze problems from the perspective of their own activity and develop a new AS.

For the following phase, we planned sessions in which each group would present to the other proposals to debate possibilities of joint activity. Ahonen and Virkkunen (2003) reported a similar strategy upon noticing two AS with shared objects and differently designed scripts.

Before the sessions with the local managers, we interviewed and observed the activities of the principal, deputy principal, and pedagogical coordinator to formulate other hypotheses on the management AS. We also video recorded three teachers along a workday and discussed their actions individually. Analysis of the collected information was meant to support the mirror data to be used in the following sessions (Virkkunen and Newnham 2013). This strategy also served to compensate for the scarcity of data collected at CL onset.

Other problems solved concerned the choice of that particular school as unit of intervention, even though the local management had not acknowledged the need for it, nor had they negotiated the intervention method with us. These two circumstances accounted for their mistrust toward CL and led them to ask why their school had been the single one to be included.

We also realized that, in Brazil, the term "intervention" might denote mandatory, top-down changes and evoke, for instance, memories of the military dictatorship. It seems to us this was one of the problems associated with CL, especially in the principal's view, as she said she was feeling somehow persecuted, because the "school was under intervention."

To clarify CL object and the characteristics of expansive learning, the ninth session was fully devoted to joint analysis of intervention. Once the problems in the relationship between teachers and management became clear, the strategy to attenuate the feelings of persecution and promote participation was to invite the managers to the bimonthly meetings of the researchers-interventionists with the central management. Joint evaluation helped clarify some of the method characteristics to the participants; as a result, questioning shifted to the governance of changes and the relationship among participants.

The sessions conducted with the teachers alone served to develop agency, particularly concerning the new analysis and new activity potential. The teachers were able to suggest other modes of collaboration also with the local management. Nevertheless, analyzing negative aspects of their work was uncomfortable to the managers, which was the direct trigger for the discontinuation of CL. What seems to have transpired is precisely what Virkkunen and Newnham (2013, p. 109) observed:

> Encountering negative information about one's activity is never easy and the participants often deny the factuality and significance of the evidence of problems present in the mirror data or find justifications for the situation thereby maintaining the current practice.

6.5 Final Considerations: Lessons Learned by Researcher-Interventionists

The present was one of the first CLs conducted by our group. Follow-up in group discussions, particularly Intervention Study Seminars (see Sect. 15.2), resulted in several essential methodological lessons for the planning and execution of other interventions described in this book.

The first lesson learned concerns the relevance of negotiation, which is not emphasized in the literature we reviewed. In the present case, negotiation did not take place at the local level, which resulted in resistance to joint proposals for change. As negotiation involved the central management only, the local managers did not acknowledge the need for intervention.

Based on the comparison of the present with other experiences of our group, we conclude that, in Brazil, negotiation vis-à-vis formative intervention is necessary to develop shared objects. It is much more difficult to solve conflicts within and between AS without the collaborative participation of managers. When managers do not achieve the expected learning, they resist and/or discontinue the process, as occurred in the present case. Centralizing and authoritarian management models still prevail in Brazil, while demands for intervention do not necessarily originate within the organizations facing problems. Therefore, negotiation should be considered as the first phase of CL (see Fig. 2.5).

The principal investigator's needs—derived from deadlines set at her graduate course—had the highest weight in the formulation of the demand and planning of the process, which resulted in challenges which had to be overcome at a later time. The main lesson learned in this regard is that the formulation of adequate hypotheses and accurate understanding of the history and contradictions of an activity and the corresponding solutions—and consequently, better planning of interventions—are closely related to ethnographic data collection. This is to say, the lack of commitment to achieve thorough acquaintance with the target activity before the onset of sessions might substantially impair the development of formative interventions. The fact we lacked a general overview, even if hypothetical, of the analyzed activity probably influenced the challenges we met in later sessions upon attempting to make proper room for dialogue among the participants.

The formulation of the demand further evidenced dissociation between the teaching-learning-outreach goals and the aims of intervention. In our milieu, interventions seldom derive from the needs of organizations but of investigators. They are not shared interests, which makes acknowledging the need for change difficult.

Another methodological lesson derives from the fact sessions lasted just 1 h, which interfered with the development of expansive learning actions; Virkkunen and Newnham (2013) recommend 2-h sessions. In our opinion, shorter sessions are insufficient to induce microcycles of expansive learning. In the present and other interventions performed by our group, we noticed that participants need some time to talk "freely." Time is also needed to conclude double stimulation in one single session. Then, later sessions never start exactly at the point where the previous one ended, but researchers-interventionists are compelled to "retrace some steps." We compensated for the weaknesses of short duration through a larger number of sessions. However, since the sessions were scheduled at times allocated to training and planning of teaching activities, we possibly interfered with them, which displeased the managers.

In the present study, we described the conflict between the three activities as a sort of "sabotage" of CL. This might have been the result of the management model in force, the various coexisting scripts, and the expected patterns of participation. We mentioned the cases of the daily report and the weekly plan to discuss modes of collaboration and disturbances. We further addressed the pedagogical mediation developed by the managers based on historical changes. There is a close relationship between these instruments established by the authoritarian management model—expressed as control over teachers—which runs against the autonomy needed in creative work.

Managers should support teachers and thus behave as mediators to facilitate and potentiate the pedagogical process in the classroom. The detected disturbances indicated that support actions did not exactly as expected, and the discomfort triggered by CL probably was a manifestation of a primary contradiction within the division of labor: a vertical, authoritarian, and control-centered versus a horizontal and democratic division of labor.

Primary contradictions are the most difficult to solve (Engeström 2015) because they are part of the essence of the capitalist society, and thus any transformation demands long-term sociocultural changes. A focus on the complexity and social shortcomings of teaching work in Brazil shows that the required quality and creativity depend on a more horizontal and democratic division of labor. Within such context, control-centered management represents a hindrance to expansive learning.

Expanding the activity object, or developing new ways to share it, demands changing the nature of the power relations between the management and the teachers' activity. Daniellou et al. (2011) call the attention to the relevance of more democratic management for the safety and well-being of workers. However, one might assume that neither investigators nor local actors are strong enough to transform the authoritarian nature of the division of labor.

This study was financed by the São Paulo Research Foundation with a fellowship for postdoctoral stage of Amanda Aparecida Silva Macaia (FAPESP 2015/13301–4).

References

Ahonen, H., & Virkkunen, J. (2003). Shared challenges for learning: Dialogue between management and front-line workers in knowledge management. *International Journal of Information Technology and Management, 2*(1,2), 59–84. https://doi.org/10.1504/IJITM.2003.002449.

Assunção, A. A., & Oliveira, D. A. (2009). Intensificação do trabalho e saúde dos professores. *Education and Society, 30*(107), 349–372. https://doi.org/10.1590/S0101-73302009000200003.

Avellar, E. T. (2017). *O método "Laboratório de Mudança" como intervenção formativa em uma escola do ensino fundamental: uma nova perspectiva em saúde do trabalhador [thesis]*. São Paulo: Universidade de São Paulo. Portuguese. https://doi.org/10.11606/T.6.2017.tde-10072017-151939.

Daniellou, F., Simard, M., & Boissières, I. (2011). *Human and organizational factors of safety: A state of the art*. Toulouse: Les Cahiers de la Sécurité Industrielle. FONCSI.

Engeström, Y. (2008). *From teams to knots: Activity-theoretical studies of collaboration and learning at work*. Cambridge: Cambridge University Press.

Engeström, Y. (2015). *Learning by expanding: An activity-theoretical approach to developmental*. New York: Cambridge University Press.

Engeström, Y., Virkkunen, J., Helle, M., Pihlaja, J., & Poikela, R. (1996). The change laboratory as a tool for transforming work. *Lifelong Learning in Europe, 1*(2), 10–17.

Fichtner, B. (1984). Coordination, cooperation and communication in the formation of theoretical concepts in instruction. In: Hedegaard, M., Hakkarainen, P., & Engeström, Y. Learning and teaching on a scientific basis: Methodological and epistemological aspects of the activity theory of learning and teaching (pp. 207–227). Aarhus, Denmark: Aarhus University Press.

Garcia, M. M. A., & Anadon, S. B. (2009). Reforma educacional, intensificação e auto-intensificação do trabalho docente. *Education and Society, 30*(106), 63–85. https://doi.org/10.1590/S0101-73302009000100004.

Haapasaari, A., Engeström, Y., & Kerosuo, H. (2016). The emergence of learners' transformative agency in a Change Laboratory intervention. *Journal of Education and Work, 29*(2), 232–262. https://doi.org/10.1080/13639080.2014.900168.

IPEA – Instituto de Pesquisa Econômica Aplicada. (2000). *Texto para discussão n° 754. A reestruturação das políticas federais para o ensino fundamental: descentralização e novos mecanismos de gestão*. Brasília: IPEA.

Oliveira, D. A. (2007). Política educacional e a re-estruturação do trabalho docente: reflexões sobre o contexto latino-americano. *Education and Society, 28*(99), 355–375. https://doi.org/10.1590/S0101-73302007000200004.

Raeithel, A. (1983). Tätigkeit, Arbeit und Praxis. Frankfurt am Main: Campus.

Virkkunen, J., & Newnham, D. S. (2013). *The change laboratory: A tool for collaborative development of work and education*. Rotterdam: Sense Publishers.

Chapter 7
Open Doors, No Slots: Application of the Change Laboratory at the Renal Replacement Therapy Unit of a Public Hospital

Cristiane Parisoto Masiero, Ildeberto Muniz de Almeida, Marco Antonio Pereira Querol, and Rafael Junqueira Buralli

Abstract The present chapter discusses the application of the Change Laboratory (CL) method at the hemodialysis service of a public hospital in Brazil. The initial context of the intervention was characterized by a high absenteeism rate among the nursing staff and reports of a crisis in the unit. The chapter describes an experience of the application of the method from management support to difficulties in the determination of the object of the activity. The problems detected derived from changes made in 2012, which resulted in the need for patients to stay several months at the hospital until slots for outpatient hemodialysis became available. Along this period, they received care from other hospital units. The intervention evolved along several stages, which required much learning from both researchers and staff to reconceptualize its object. Mirror data evidenced emotions which began to be shared, followed by critical analysis of contradictions. The resulting discussions enabled and encouraged initiatives for change in the organization of the activity.

C. P. Masiero (✉)
Department of Public Health, Botucatu School of Medicine, São Paulo State University "Júlio de Mesquita Filho", Botucatu, SP, Brazil

Workers' Health Reference Center, Bauru, SP, Brazil
e-mail: cparisoto@uol.com.br

I. M. de Almeida
Department of Public Health, Botucatu School of Medicine, São Paulo State University "Júlio de Mesquita Filho", Botucatu, SP, Brazil

M. A. P. Querol
Department of Agronomic Engineering, Federal University of Sergipe, São Cristóvão, SE, Brazil

R. J. Buralli
Department of Environmental Health, School of Public Health, University of São Paulo, São Paulo, SP, Brazil

R. A. G. Vilela et al. (eds.), *Collaborative Development for the Prevention of Occupational Accidents and Diseases*, https://doi.org/10.1007/978-3-030-24420-0_7

7.1 Introduction

The Change Laboratory (CL) is a formative method that involves collaboration between researcher-interventionist and practitioners to transform work practices (Engeström et al. 1996). According to this method, management support to the implementation of changes is essential to foster the transformative process (Virkkunen and Newnham 2013). The context and tools used, such as double stimulation, involve socio-cognitive processes, including emotional confrontation of problems, followed by distant intellectual analysis (Virkkunen and Newnham 2013).

The method has been applied by the researchers within the context of the present "TemATico" project since 2014, including interventions in several areas. Such interventions were developed in response to the researchers' expectations to reach beyond an accurate diagnosis established in previous projects, through the application of the activity ergonomics approach combined with accident analysis based on the Model of Analysis and Prevention of Accidents (MAPA). However, different from Finland, where the CL method was originally formulated, most interventions in Brazil are triggered by lawsuits or demands by researchers-interventionists. Few companies voluntarily develop initiatives to improve the organization of work based on participatory methodologies, i.e., in which the employees are the agents of change.

The following intervention is an example of the application of CL originated in the demands from both a group of researchers and a hospital with a participatory management model.

The present chapter discusses the relevance of management support to strengthen interventions and the path followed by researchers and workers to understand the problem from the initial demand to its redefinition along intervention. It also describes the main contradictions identified in the target activity, obstacles met, and lessons learned.

7.1.1 *Care to Chronic Renal Patients in Brazil*

In Brazil, universal healthcare is a social achievement ensured by the Unified Health System (Sistema Único de Saúde—SUS) which covers from primary to tertiary care. Medium and high complexity procedures represent about 40% of the health budget, including expenses related to renal replacement therapy (RRT) which concentrates almost 50% of the resources, with a tendency to increase (Junior and Mendes 2015). RRT is often required when the glomerular filtration rate is reduced to levels between 10 and 15%. At this point, patients are considered to exhibit end-stage renal disease (ESRD) which is a permanent and irreversible condition. To replace the renal function lost, besides kidney transplantation, which is still underutilized in Brazil, both peritoneal dialysis (PD) and hemodialysis (HD) might be used as efficient renal substitution methods (Carneiro et al. 2015).

In 2010, less than half of the patients requiring RRT in the world had access to treatment. In addition, it is estimated that the number of people requiring RRT will increase from 2.618 million in 2010 to 5.439 million in 2030, i.e., more than twice (Liyanage et al. 2015).

According to the Brazilian dialysis census published in 2016, the estimated number of patients with ESRD in 2000 was 42,695, while in 2010 it was 92,091 and 112,004 in 2014, corresponding to an increase of 262% along 14 years (Sesso et al. 2016). The most common treatment in Brazil is HD delivered at outpatient clinics or hospitals. In turn, PD is performed by the patient or a family member daily at home through a peritoneal catheter inserted into the abdomen (Daugirdas et al. 2012).

Considering the current (and future) scenario of chronic renal disease, there is an urgent need for innovative preventive models and cost-effective therapies, especially in low-income countries (Liyanage et al. 2015).

7.2 CL at the RRT Unit of a Public Hospital

The present study was conducted at a public hospital opened in 2003 at a medium-sized municipality in the central-western area of the state of São Paulo, Brazil. It is a part of the SUS tertiary healthcare network and serves a population estimated as 1.8 million individuals across 68 municipalities. The RRT unit started its activities in 2007, and ever since RRT mainly consists of HD, followed by PD in ratio of five to one, respectively. Kidney transplants are much less common and are performed at another hospital in the area.

The RRT nursing staff is subdivided into HD room team, HD acute renal insufficiency (ARI) team, and PD team. Nephrologists are distributed similarly and also perform routine visits to outpatients in the pre-dialysis stage. The HD room team provides care to 197 chronic renal patients who receive outpatient HD three times a week. The ARI team works on shift regimen (6-h daytime and 12-h night shifts) at intensive care units (ICUs) which admit acute renal patients. These two teams are responsible for about 2500 HD sessions per month. The PD team provides training and monthly follow-up to about 40 patients and their families for performance of PD at home.

7.2.1 Negotiation and Data Collection

The idea to apply the CL method at this hospital emerged during a meeting with the coordination of the Workers' Health Reference Center (CEREST) of the city of Bauru, São Paulo, Brazil, and the hospital social worker for human resources (HR), when she learned that someone was seeking an opportunity to apply a formative methodology as a part of a master's project. After this first talk, a meeting was held in May 2016 with the hospital managers, who received the proposal with much enthusiasm.

The hospital managers suggested applying CL at the RRT unit due to the many complaints against the HD team, which put much more demands, compared to other hospital staff, exhibited high absenteeism rates, and entered in frequent conflict with the unit manager.

The initial stage of the negotiations lasted about 1 month. Meetings were held between the researcher-interventionist agents, HR social worker, nurse responsible for the HD unit, and Occupational Health and Safety Department (SESMT). The social worker mediated between the hospital and the researchers, followed up, and supported the whole process of intervention. This group continued to meet periodically to discuss and plan the intervention.

During the stage of data collection, the researcher-interventionist agents analyzed absenteeism data, the list of allocated employees per shift, and the number of dialysis sessions per month, among other variables. At the same time, they visited the unit during all three shifts, when they held short briefing meetings on the CL methodology, and scheduled individual and group interviews, to a total of eight and six, respectively. The researchers also interviewed inpatients waiting for HD slots and filmed the interviews, which were later used as mirror data in CL sessions.

In addition to interviews, on-site observation of the work process was performed to achieve a better understanding of the RRT unit activity and teams.

7.2.2 Initial Demand

The initial demand pointed to problems in the interaction between the HD team and other professionals responsible for inpatients with ESRD as the focus of intervention. Actors included wards staff, ICUs where RRT was performed, the Hospital Infection Control Committee (HICC), and the hospital admission management staff.

Interviews showed that the problems actually went much beyond the HD and RRT teams. One of the reports that called our attention stated that an ICU employee had refused to let a HD colleague use the bathroom in the unit. Complaints from other units against the HD team emerged during interviews, for instance, "...they believe they are the heart of the hospital...," alluding to employees' reports stating that HD brought considerable financial resources to the hospital. The pharmacy staff manifested discomfort with their requests for immediate delivery of medications. To collect more information on possible complaints against the HD team, a questionnaire was applied at the 43 hospital departments which somehow interacted with the RRT unit. Despite the researchers' awareness of the relevance of participation, the response rate was only 9%, and no case included complaints against the HD unit.

Several factors led to the need to extend the period of data collection. The HD nurse supervisor was replaced, which was negatively received by the employees, the principal researcher travelled for a fellowship abroad, and a 52-day strike interrupted the intervention.

7.2.3 Expansion of the Study Focus and Renegotiation

In the interviews performed during the stage of ethnographic data collection, we noticed that the demand for RRT, HD in particular, exceeded the hospital capacity. The interviewed nephrologists reported that upon being requested to choose a treatment method, patients and their relatives preferred HD over PD. This preference might perhaps be justified by the fact that HD is performed at the hospital under medical supervision, while PD is performed at home. Due to long wait times, the nephrologists decided to admit the patients to the hospital until a slot became available at the HD unit, which could take from several days to months. At the time of data collection, some patients had stayed more than 100 days at the hospital. After hospitalization, care to ESRD patients was provided by several hospital departments, so these patients became an object of activity shared among different units.

This system seemed to work well in the beginning. The patients stayed at the hospital, usually at the 4th floor, and were taken by members of the nursing staff downstairs to receive HD three times a week. HD sessions occurred in an ICU room, and dialysis was performed by the ARI team, who up to then had only treated acute renal patients admitted to ICUs. However, given the continuous arrival of patients through several paths (emergency department, outpatient clinics, judge orders) and the impossibility to refuse patients for lack of beds[1]—according to the emergency and urgent care regulation known as "zero bed"[2]—the situation gradually deteriorated over time.

The interviews revealed that the origin of the problem was the hospital decision to admit an increasing number of chronic renal patients, although they exceeded the installed capacity for HD considering all three shifts. A decision was then made to admit the patients to wards for other specialties and transport them to ICU on the days scheduled to receive HD. This change led to new interactions and also to complaints from the employees in other departments and the patients subjected to too long stays at hospital. After the end of data collection and analysis, the hypothesis which guided the present study was that the lack of HD slots and the decision to admit waiting patients led to new interactions among the hospital staff which caused the problems detected (see Fig. 7.1).

---

[1] The situation in which a hospital is overloaded to the point it judges it cannot safely accept further patients.

[2] Situation considered in the urgent care regulations, according to which the physician charged of transferals does not accept unavailability of beds as a reason to deny care to urgent patients.

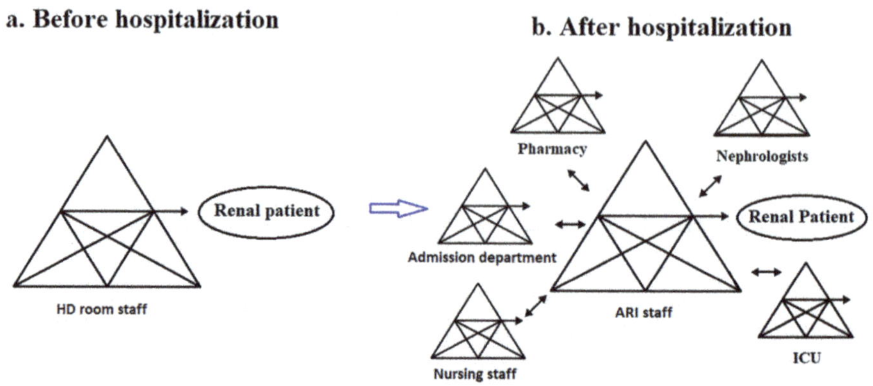

Fig. 7.1 Activity system before (**a**) and after (**b**) the decision to admit chronic renal patients. (Adapted from Engeström 2015)

7.2.4 Intervention Planning

We thus understood this was a case of a boundary crossing CL, which demanded improving communication and collaboration between two or more activities sharing the same object, analyzing the causes of ruptures and disturbances jointly, and searching for new notions and tools to manage collaboration or design new arrangements for the organization of work (Virkkunen and Newnham 2013).

The work plan was formulated in a way that the first stimulus corresponded to the problems experienced in the workplace as a function of disturbances in the activities. Following Virkkunen and Newnham (2013), while the object of the different activities was the same (i.e., shared patients), it was interpreted differently by various actors involved in different activities, who had started sharing the patients after the decision to admit them to the hospital. Given the disturbances that ensued, it was necessary to cross borders in this case. Still according to Virkkunen and Newnham (2013), the solutions suggested in a boundary crossing CL might behave as secondary stimuli, and a new model might be formulated as second instrumental stimulus aiming at collaborative reorganization.

Following the redefinition of the problem, the researcher-interventionist agents met with the hospital management to present the new study hypothesis; the hospital director accepted and supported the new proposal. Next, the group which would participate in CL sessions was selected, considering the 43 departments which interacted with the RRT unit and RRT employees, including HD room, ARI and PD nurses, nursing technicians, and nephrologists allocated to the HD room. Initially, the group comprised 23 participants.

Four CEREST members composed a follow-up and planning team, including the principal researcher, a physical therapist, a nurse, and a psychologist. Participants from the hospital staff were a social worker from the HR department, a nursing technician who had previously worked at the HD unit and was currently allocated to the hospital occupational health and safety department, a psychologist from the HD unit, and the hospital general nursing manager. This group met weekly to plan the CL sessions, to a total of eight, followed by one further meeting with the hospital board to present the results.

At the first preparatory meeting, the hospital executive director reinforced their support to the initiative to collectively develop suggestions for changes at the HD unit, as well to the application of CL at the hospital:

> I believe this is an excellent proposal, a different tool (...) I believe that the most important thing of the method as such, of sessions, is that we get to talk about what we truly think. The impressions we have about the problem. Because sometimes we don't want to talk, because someone said something, but I have a different idea, and therefore what is the point? Precisely, the confrontation of ideas, of thoughts, so we can rebuild and try to find solutions together (…) It's an excellent opportunity to find the solution we haven't found yet (…) If we think together we'll be successful!
>
> (Hospital executive director).

7.2.5 CL Sessions

As the CL method had been presented in a previous meeting, the first session began with a brief introduction to the CL method, and then the groups were allowed room to express their doubts about the purpose of intervention, which for them ought to be much more focused on the relationship between teams. Before addressing the historical origin of the problems, the group observed that the solution was to increase the number of HD machines, which would automatically make more slots available. However, this suggestion was beyond the group's prerogatives:

> Lack of slots (…) this is something we won't be able to solve now (…) I had understood that our focus would be on more on the staff (…) This is something we can't solve with the Change Laboratory, it doesn't depend on us… and that would be the focus? (Nursing supervisor)
>
> …But we'll work on the repercussion of this in the various areas; "ah, nursing is overload, then, how it is?" Now the issue of patients, slots, this falls out of our control… (Nephrologist 1)
>
> Does it only depend on the creation of new slots? (Researcher)
>
> Yes, absolutely. On the machines (Several participants)
>
> If we get into the issue of slots, I believe we'd do nothing but chase our tails ... This what I feel. (Nursing supervisor)

At this point, the researchers showed the group the image of an iceberg, to convey the idea that the aim of CL sessions is for participants to forsake the visible part on the surface and to look deeper into the "hidden" aspects of problems. In this case, the participants are encouraged to look for other solutions or improve the activity, which could go beyond the creation of new HD slots (Fig. 7.2).

Still during the first session, the researchers presented as first stimulus a fragment of narrative of a patient, who had stayed 4 months at the hospital waiting for an outpatient HD slot:

Fig. 7.2 First CL session at the hospital

> When you're admitted to the hospital, you believe you'll stay 10, 15 days. But then you realize that one month has passed, two, three months. I stayed 110 days (…) my routine was the same as in prison. They put you there, and you can't get out. It's like prison, just without cells… (Patient 1, 37 years old)

The patient's narrative triggered reactions, mainly among the nephrologists, who sought to justify the need for treatment and the lack of outpatient slots. Nonadherence of patients and relatives to PD was mentioned as a problem to be overcome to reduce the waiting list for HD. Seemingly, some unsuccessful attempts at increasing PD had been made, while the preference for HD was attributed to the fact that it is performed under medical supervision.

Next, the representatives of teams were asked to express their views on the impact of the admission of chronic renal patients on the work routine and to suggest solutions or improvements. The areas and groups that reported the highest impact were:

- The 4th floor, where patients were admitted and received medical and nursing care. The nursing team transported the patients to the ground floor for the weekly HD sessions.
- ICU, located on the ground floor, where a room was adapted for the HD machines; care was provided by the ARI team.
- The ARI team.
- The nephrologists, who complained of a greater workload due to the increasing number of admissions.
- The HICC representatives pointed to the higher susceptibility of renal inpatients to acquire severe hospital infections as a function of the venous and arterial access for dialysis.
- The admission control department rated the long occupancy of beds as a problem, given the current situation of lack of beds and waiting lists in the emergency departments; the ongoing occupancy rate was 95%, and on occasions, it had reached 98%.

The responsibility for the care of chronic renal inpatients was thoroughly discussed in the sessions. No team considered that the responsibility for patient care was theirs. One of the physicians stated he had already followed up renal inpatients during his shift, even when he felt it was pointless, since they usually did not need care, but was merely waiting for a HD slot. According to him, the responsibility for these patients corresponded to the nephrology service. This was also the opinion of the ICU staff, which observed they had simply lent a room for HD. In turn, the ARI nursing technicians performed HD in the ICU room, isolated from the nurses and nephrologists at the HD room. They saw themselves as providing a specialized service, a situation experienced in others ICUs routine, where patients received all care, as e.g., change of dressings.

The problems between teams became even worse in the case of unexpected events, such as cardiac arrest during dialysis at ICU. As a rule, the ICU staff provided urgent care. However, the participants saw some situations as a cause of conflict, for instance, replacement of crash cart supplies, procedures such as dressings, and transport of dead patients to the morgue, among others. None of the two teams considered these activities a part of their job description. Shortly, the participants manifested difficulty to understand who was responsible for these patients:

> ... There are two doctors at ICU… but none of them is responsible for dialysis (…) the nephrologist is in the room… and some emergency occurs in the room, and you can't leave… and the guy at the ICU, no, no… ICU has no responsibility… ICU is ICU… a closed unit Then who's responsible for that patient?... No one… (HICC physician)
>
> Because no one is responsible for this patient ... (ARI Nurse)

7.2.6 *Historical Analysis of the RRT Unit*

The RRT unit was created at the analyzed hospital in 2006, in the same room where ICU patients currently receive dialysis as a result of the lack of slots. In 2007, the HD unit moved to its own premises, and the ICU room was deactivated. Until 2012, the RRT unit expanded as a function of the demand. Shifts increased from two to three and new HD machines were installed. The first patient waiting for a HD slot was admitted in 2012. At the same time, the ICU room was allocated to the treatment of inpatients waiting for outpatient HD slots.

At that time, an agreement was made by which ICU would lend a room for HD, the ICU staff would be charged of emergency care, and nephrologists on duty would provide care as needed. In the beginning, this arrangement seemed to work well, but as the number of ICU beds and admissions of chronic renal patients increased, also did the workload of all the involved professionals.

The HD room staff remained stable, while the waiting list increased, as the number of available slots depends on the number of HD machines.

As a result, the model chosen to deal with the increasing demand began to give signs of exhaustion. The arrival of new patients requiring dialysis had direct impact on the ARI nursing technicians, as their workload increased. For the nephrologists, with the initial arrangement, working during weekends and holidays was financially interesting. Nephrologists on duty were charged of prescribing dialysis to inpatients, follow them up, and provide care in case of unexpected events. However, when the number of patients increased, they began to complaint, to the point the weekend shifts became known as "shifts of utter unhappiness."

The plan for the fourth session was to perform a historical summary of the hemodialysis unit, but the absence of the nephrologist and other professionals served as a warning signal to the researchers. For this reason, the fifth session was devoted to receive feedback on the development of the CL until that point. With the help of the hospital psychologist, the planning group prepared an activity to encourage the participants to discuss their impressions of the previous sessions. On this occasion, the participants stated that the open discussions of problems made them feel uncomfortable and exposed. This was mainly the perception of the nephrologists, who further stated they felt frustrated and guilty in regard to the situation, which they saw as with "no way out" or of a double bind, as on the one hand they had to admit patients with ESRD, but on the other there were no slots for their treatment. This perception corresponds to a well-known situation in formative interventions, such as CL, even more accentuated in the Brazilian context as a function of the lack of room for discussion and collective creation in the workplace.

Also in the fifth session, the researchers suggested forming a subgroup for the following sessions, with an eye on the RRT team, given the high number of participants absent in the previous session and the detachment of other participants, who possibly did not see themselves as having a direct connection with the object of the activity. To facilitate participation given the hectic pace of work at the institution, a strategic change of venue was made, and the sessions, originally held at the hospital boardroom, moved to a room in the RRT unit.

Following this move, the now reduced group attempted to formulate a new activity model, which however still had narrow focus on the isolated elements of the activity system, and included proposals such as reducing the ARI team workload through new hires and training of other nursing teams.

For the seventh session, the mirror data presented corresponded to the flow of chronic renal patients cared at the nephrology outpatient clinic from the pre-dialysis stage to kidney failure and need for RRT. This presentation allowed the participants to realize that the patients received the diagnosis of ESRD and were requested to choose between HD and PD on one and the same visit. This led the group to suggest the intervention of a multidisciplinary team in the pre-dialysis stage to provide integral care and elucidate patients on their therapeutic options. The expectation was that this approach would increase the adherence to DP and reduce the demand for HD. The same approach would be followed with inpatients to reduce the number of admissions. The group designed a new care delivery model, which was presented at the eighth session to be thoroughly discussed in the following ones, but they decided to get the approval of the hospital board before continuing. The group also began to discuss preventive actions likely to reduce the demand in the medium and long run, but the suggestions were not taken forward.

The proposals collectively formulated by the group were approved by the hospital management. The latter emphasized the relevance of the discussions in the search for solutions for the consequences of the admission of chronic renal patients, among which unnecessary bed occupancy. This fact was particularly significant, given the lack of hospital beds in the area, in addition to the high cost for the public health system.

After this meeting, which took place in mid-December 2017, the hospital staff suggested scheduling the following ones after January, since many employees would be on vacation that month. However, several changes occurred in the meantime:

- A reform was planned for the 4th floor to enable dialysis on the premises and thus avoid the need to transport patients to the ground floor (ICU). However, this project was never implemented.
- Both nephrologists in charge of the HD unit since its creation resigned.
- The hospital hired a nephrologist with experience in PD as supervisor of the RRT unit. The expectation was to increase PD through the use of softer technology (Merhy 2002)[3] which allows performing dialysis at home, with consequent decrease in the HD demand and elimination of the need to admit patients to the hospital.

The DP option for RRT is largely used at the teaching hospital of a medical school at a reference municipality 100 km away from the hospital where the present study was conducted. The hospital staff took the initiative to learn how this teaching hospital works with less resources and serves a larger number of patients needing RRT. Benchmarking for the following CL sessions was performed by the team itself, being thus considered a manifestation of transformative agency.

Currently, the activity at the HD unit is being reconceptualized, which involves changes in the care delivery protocols (new rules) and enhancement of PD as an alternative to HD through agreements to purchase equipment (new tools). As a function of the patients' adherence to PD, by September 2018, there was absolutely no chronic renal patient admitted to the hospital or in the waiting list for dialysis.

While these changes had been mentioned as options in CL sessions, they were implemented by the hospital management in the interval between sessions at the turn of the year.

The management and planning group observed that the solutions then being implemented could not have been possible without the diagnosis established in the CL sessions and expressed the desire to apply the method to other hospital areas. As a result of this desire, in May 2018, a CL training workshop took place at the hospital, practically without the participation of any professional from the RRT unit, reflecting their moment of changes, but with a large number of hospital employees from other units.

[3] Merhy classifies the work process in healthcare according to the use of hard, soft-hard, and hard technologies. Hard technologies concern instruments (e.g., machines), hard-soft technologies technical expertise, and soft technologies the relationship between individuals, associated with actual work and embodied in actions.

7.3 Discussion

The CL at the analyzed hospital was conducted among symptoms of disturbances in the activity (absenteeism and mutual accusations of non-responsibility for the chronic renal patients) which were shared in the sessions. Such manifestations of emotions considerably contributed to evidence contradictions, in this case, mandatory hospital admission of patients waiting for HD slots. A similar situation was identified by Kerosuo in a CL intervention at a university hospital in Finland, when operating rooms were closed and the teams were left with no choice, which resulted in tension and crisis. Sharing emotions followed by critical analysis allowed the group evolve in the search for a diagnosis and solutions (Kerosuo 2006; Hannele 2011).

In the CL described here, the sessions allowed for emotional confrontation, which led to better and more thorough understanding and visualization of the problem. The discussions enabled an expansion of the comprehension of the object of the activity, i.e., the chronic renal patients, the responsibility for whom had been initially attributed exclusively to the nephrology/RRT staff. In turn, after the decision was made to admit these patients, care began to be provided by multiple teams, while no one assumed the responsibility. The initial agreement for shared care gradually eroded as a function of the increase of inpatients, multiple demands, and unexpected events. Already charged of multiple tasks, the teams opposed resistance, which conspired against the strategy of admitting renal patients, resulting in a vicious circle. The discussions led to a new understanding, to wit, that the admission of patients with ESRD caused new interactions and imbalance between the components of the hospital activity system. The initial agreement for adjustments and responses proved to be useless to solve problems, while the actors insisted in following the preset path. Thus emerged the conditions to provide solutions focusing on the main detected problem.

According to Engeström (1995), object reconceptualization in CL applied to the healthcare field is characterized by a constant movement of reconstruction, either by new technologies or new knowledge on medical diagnosis.

One of the difficulties identified in the present study derived from the short time available for CL sessions at the hospital. The work schedule hindered, for instance, the participation of one and the same ARI nursing technician in all the sessions, for which reason the technicians took turns to represent their group. This turnover hindered the progression of sessions, as topics had to be resumed with each new participant, which was tiresome for the ones who had participated from the beginning. In addition, since sessions took place during the working hours, some technicians resented the fact that some colleagues did not recognize them as their representatives but as "lazy people who do not want to do their job."

These difficulties notwithstanding, the group did indeed learn much, as manifested through changes in activity components to improve the working conditions, as well as in the activity as such, as evidenced by changes in the protocol for patient care. Contradictions have a power to trigger change through expansive learning (Engeström 1995). Although they were mentioned here, they are not the focus of the present chapter.

7.4 Final Considerations

In the present chapter, intended to report the application of CL at the RRT unit of a Brazilian hospital, we approached the relevance of management support to a formative intervention, as well as the cycle of expansive learning of the hospital staff and researcher-interventionist agents concerning the determination of the activity object.

The hospital management supported the intervention. Urgent needs drove the group look for solutions, including object reconceptualization and understanding of the historical events that had led to a state of crisis.

Understanding the nature and origin of the problems was crucial in the search for lasting changes, and sustained the process of expansive transformation.

The main aspect of expansive learning within this context might be associated with the perception that care delivery is the product of interactions among multiple actors, including healthcare providers and patients. The changes implemented at the hospital gave rise to several problems in care delivery to ESRD patients, which could not be solved by merely replacing the actors.

In our view, no matter how well interventions are planned, they are only partially predictable, because each system has a dynamics of its own, resulting in chance outcomes unforeseeable to actors or intervention agents. Changes are always complex and multifaceted social constructions, where different flows of events and interests are intertwined and demand permanent negotiation and construction (Long 2001).

All difficulties notwithstanding, learning occurred continuously all along the process of problem analysis and planning of possible future activities. The CL provided the group a comprehension of a proposal of collective construction, although it still meets resistance as a function of the traditional decision-making model adopted in Brazilian institutions. In spite of the relevance of the diagnosis of the core problem, we recommend continuing the study to evaluate the adequacy of the implemented changes, as well as new interactions and problems.

Special Thanks We would like to thank all our colleagues from CEREST, especially Marcia, Roseli, Cesar, and Natasha, for all the support during the CL intervention.

We also extend special thanks to patients and staff from Hospital Estadual de Bauru, as well as the hospital's management and HR.

Thank you!

References

Carneiro, É. C. R. D. L., Pecoits, R. F. S., Ribeiro, S. C., Pestana, J. O. M. D. A., Polito, M. G., Costa, M. H. S. N., et al. (2015). *Manejo clínico das doenças renais*. São Luis: UNASUS-UFMA.

Daugirdas, J. T., Blake, P. G., & Ing, T. S. (2012). *Handbook of dialysis*. Lippincott: Williams & Wilkins.

Engeström, Y. (1995). Objects, contradictions and collaboration in medical cognition: An activity-theoretical perspective. *Artificial Intelligence in Medicine, 7*(5), 395–412.

Engeström, Y. (2015). *Learning by expanding: An activity-theoretical approach to developmental research*. New York: Cambridge University Press.

Engeström, Y., Virkkunen, J., Helle, M., Pihlaja, J., & Poikela, R. (1996). The Change Laboratory as a tool for transforming work. *Lifelong Learning in Europe., 1*(2), 10–17.

Hannele, K. (2011). Caught between a rock and a hard place: From individually experienced double binds to collaborative change in surgery. *Journal of Change Management., 24*(3), 388–399. https://doi.org/10.1108/09534811111132767.

Junior, A. P., & Mendes, A. N. (2015). O Fundo Nacional de Saúde e a prioridade da média e alta complexidade. *Argumentum., 7*(2), 161–177.

Kerosuo, H. (2006). *Boundaries in action. An activity-theoretical study of development, learning and change in health care for patients with multiple and chronic illnesses*. Helsinki: Helsinki University Press.

Liyanage, T., Ninomiya, T., Jha, V., Neal, B., Patrice, H. M., Okpechi, I., et al. (2015). Worldwide access to treatment for end-stage kidney disease: A systematic review. *The Lancet, 385*(9981), 1975–1982. https://doi.org/10.1016/S0140-6736(14)61601-9.

Long, N. (2001). Development sociology: Actor perspectives. In *The case for an actor-oriented sociology of development* (pp. 9–29). London: Taylor & Francis.

Merhy, E. E. (2002). *Saúde: Cartografia do trabalho vivo*. São Paulo: Editora Hucitec.

Sesso, R. C., Lopes, A. A., Thomé, F. S., Lugon, J. R., & Martins, C. T. (2016). Brazilian chronic dialysis census 2014. *Jornal Brasileiro de Nefrologia, 38*(1), 54–61. https://doi.org/10.5935/0101-2800.20170049.

Virkkunen, J., & Newnham, D. S. (2013). *The Change Laboratory: A tool for collaborative development of work and education*. Rotterdam: Sense Publishers.

Chapter 8
Envisioning a Solution for a Runaway Object: A Formative Intervention in a Child Labor Combat Network

Sandra Donatelli, Rodolfo Andrade de Gouveia Vilela, Marco Antonio Pereira Querol, and Sandra Francisca Bezerra Gemma

Abstract Technological changes and increasing international competition in jewelry and semiprecious jewelry manufacturing, which took place in recent decades, led manufacturers in Limeira, São Paulo, Brazil, to implement outsourcing strategies which promote child exploitation. Despite the efforts of regulatory agencies, child labor could not be eradicated to this day. As a fact, the pattern of spread of child labor makes it uncontrollable and thus what Engeström calls a runaway object. Such objects pose a considerable challenge, because they lack a center for control, but combat depends on the cooperation of various interdependent activities. The aim of the present chapter is to analyze Change Laboratory (CL) sessions to describe the learning trajectory of Limeira Municipal Committee for Eradication of Child Labor and Adolescent Work Protection (COMETIL). Ever since its creation, COMETIL met difficulties to develop an interinstitutional network of public services targeting child and adolescent protection. Intervention served to consolidate COMETIL and define a viable object within such diffuse and complex set of activities. For analysis, we had resource to notions such as runaway object, relational agency, and infrastructure. We conclude that the object develops as COMETIL actors' agency escapes the rigid structure of the various institutions to enable the provision of assistance to families with children who work.

S. Donatelli (✉)
Occupational Safety and Medicine Foundation Jorge Duprat Figueiredo – FUNDACENTRO, São Paulo, SP, Brazil
e-mail: donatelli@fundacentro.gov.br

R. A. G. Vilela
Department of Environmental Health, School of Public Health, University of São Paulo, São Paulo, SP, Brazil

M. A. P. Querol
Department of Agronomic Engineering, Federal University of Sergipe, São Cristóvão, SE, Brazil

S. F. B. Gemma
School of Applied Sciences, University of Campinas, Limeira, SP, Brazil

R. A. G. Vilela et al. (eds.), *Collaborative Development for the Prevention of Occupational Accidents and Diseases*, https://doi.org/10.1007/978-3-030-24420-0_8

8.1 Introduction

Recent changes in production activities within the context of globalization have potential to give rise to complex, hard-to-control problems, which solution demands the collaboration of many interdependent activities which do not usually cooperate. This is the case of child labor combat in Limeira, São Paulo, Brazil. Changes in jewelry and semiprecious jewelry manufacturing posed new action challenges to regulatory agencies demanding integral actions and cooperation among them.

Social and environmental problems, such as child labor, might be characterized as runaway objects. This notion points to something which influence has potential to expand and escalate to a global scale (Engeström 2006, 2009). The Linux operating system, drug traffic, and even diseases such as cancer are well-known examples.

According to Engeström (2009), runaway objects exhibit several characteristics, such as:

- They cannot be controlled by anyone.
- They have unexpected and far-reaching effects.
- While they generate resistance, they might also trigger radically new possibilities of development.
- They are considered "natural forces" (diseases, environmental threats) or technological innovations.
- They usually begin as small problems or marginal innovations and often remain dormant or invisible for a long period of time until they break out as a crisis or breakthrough.
- They are associated with many activity systems, and their boundaries are hard to define.
- The activity might continuously change and give rise to new objects, although not necessarily focusing on the solution of problems.

This significant notion might help us understand how a problematic and shared object was reconceptualized based on collaborative agency, which emerges as a delicate and invisible tissue, but needs a stable and sound infrastructure to expand.

According to Star and Ruhleder (1996), an infrastructure develops when local or global tension is dispelled. When local practices enabled by technology might be performed in a natural manner, runaway objects determine the need of infrastructure to enable the functioning of the organization. From this point of view, infrastructure is understood as something which provides support so that relationships can occur.

The present study discusses how an infrastructure which facilitates learning to solve socioenvironmental problems characterized as runaway objects might be created. We argue that formative methods inspired in the CL method (Engeström et al. 1996; Virkkunen and Newnham 2013) might favor this type of learning.

The aim of this study was to elucidate whether formative interventions might contribute to the understanding of child labor. And if they do, how? What are the mechanisms or tools used by actors to understand the problems which emerge from networks which deal with complex and undefined objects?

For this purpose, we analyzed formative intervention sessions conducted with Limeira Municipal Committee for Eradication of Child Labor and Children Protection (COMETIL). To be sure, this study was motivated by a demand by COMETIL itself, as a way to address its difficulty to create an interinstitutional network of public services targeting child and adolescent protection. The formative intervention—based on the understanding of different activity systems (AS) within COMETIL—sought to improve the cohesion of its practices, as they target an object partially shared by municipal and state public institutions.

The present chapter begins by a brief description of COMETIL and the historical background which led to its creation. Next, we describe the methods used for data collection and analysis to continue with the results of the knowledge gained in each session. Finally, we discuss whether and how the concept of child labor, as COMETIL's object, underwent transformations leading to the formulation of collaborative solutions.

8.2 Historical Background of COMETIL and Protection of Children Who Work

Limeira, a medium-sized city in the state of São Paulo, consolidated as an important center of jewelry and semiprecious jewelry manufacturing starting in the mid-1990s. However, the development of this economic activity was associated with the emergence of child labor and increasing informal production, with small enterprises being set in the backyard by families who had lost formal jobs.

Ferreira (2005) made child labor visible in a study conducted in public schools, which confirmed that many children and adolescents participated in jewelry manufacturing at home as a result of unrestrained outsourcing. This study further found that the sewerage system was contaminated by metals used in home galvanic baths (Ferreira 2005; Vilela and Ferreira 2008).

This was the trigger for the Bank and Financial Institution Employees Trade Union of Limeira and the Workers' Health Reference Center (CEREST) of Piracicaba to start exerting pressure on the municipal government. Their goal was to promote cooperation among various sectors to deal with child labor in Limeira (Lacorte et al. 2013; Vilela and Ferreira 2008; Ferreira 2005).

The study by Lacorte (2012) evidenced the need for a network to implement policies targeting child labor aiming at reducing socioenvironmental problems derived from jewelry manufacturing.

Within this context, Limeira COMETIL—which had been created in 2003 in response to the Federal Program for Elimination of Child Labor (PETI)—was strengthened and restructured and began coordinating a group of representatives of several public and civil society institutions to implement policies able to achieve eradication of child labor, integral protection of children, and readjustment of the production process in jewelry manufacturing to eliminate outsourcing to the home environment (Lacorte 2012).

Representatives from public education, health, social development, and labor and income institutions and of employers and employees in the jewelry industry and social control organizations were mobilized to design the new structure of COMETIL. As a result, COMETIL would operate from the health surveillance perspective and lead the interorganizational network (Lacorte et al. 2013).

The Child and Adolescent Statute (ECA) bans dangerous work activities to youths under 18 years old. However, by the time of COMETIL restructuration, the social assistance center (CEPRESOM) was developing a program called "Bright Youths." The aim of this program was to provide training to adolescents under 16 years old in jewelry manufacturing procedures, such as cold or hot welding, which involves the use of dangerous acids. Shortly, the program opposed the legislation in force.

In 2008, COMETIL sought a partnership with the Labor Prosecution Office (MPT) which resulted in a Behavior Adjustment Agreement (TAC) signed with Limeira City government. This agreement made mandatory for the local, mainly the municipal, government to develop public policies aiming at eradicating child labor and readjusting the production chain in jewelry manufacturing.

TAC indicated several actions which are developed to this day:

- Creation of a training course for agents engaged in child labor eradication
- Reporting system for children and adolescents who work and receive care from the municipal health network
- Information system based on "child labor identification cards" developed by the Municipal Secretariat of Education and the municipal workers' health program (PST)

These initiatives notwithstanding, COMETIL was still opposed by the civil society, employers in particular, who denied the existence of child labor in their production chain. The limitations to the development and consolidation of the interorganizational network include two aspects: the complexity of the problem, which involves uncontrolled child domestic labor, and the fact that the representatives of both employees and employers do not adhere to COMETIL actions.

Several examples illustrate the complexity of the problem:

- Domestic work is widely spread and hard to locate.
- The low-income population depends on this type of work for survival.
- The community sees child labor and work at home as natural.
- Public education and health institutions do not consider this problem to be a priority.
- Employers deny the existence of child labor, and the ones who admit it claim it is not within the formal but the informal production chain and that the resulting negative reputation is harmful to them.

- Adherence to campaigns launched by public institutions involved with COMETIL is low, as raising awareness has low impact on the specific living conditions of the target population, since policies to develop alternative sources of income were not implemented.

Within this context of impasse, the COMETIL coordination requested an intervention from our research group to include all involved actors in a discussion of the detected problems. Intervention began to be prepared in 2016; the Committee won a simultaneous victory with the publication of Administrative Regulation No. 880 and Decree 339 which granted official status to COMETIL.

The expectation was for the Committee to develop as an interorganizational network of reflection and negotiation, essential for institutions and enterprises to look for possibilities to overcome the problem posed by child labor in the home setting.

8.3 Methods

8.3.1 Data Collection

Data were collected by means of interviews with COMETIL members working at various public institutions and civil society areas. Interviews were designed to obtain information on the institutions where the respondents worked and understand the problem posed by child labor in the home setting as COMETIL's object and institutional initiatives to combat this problem.

We interviewed 31 COMETIL members, 28 from public institution staffs, 2 trade union representatives, and 1 representative of employers. They included social workers, pedagogues, nurses, psychologists, physical educators, and engineers, among others. We also obtained information from COMETIL monthly meetings and visits to the commission, the National Service for Commercial Education (SENAC), and the jewelry trade union.

The interviewees participated in seven CL sessions, however, not necessarily in all of them. The first five sessions were conducted in 2016, and two follow-up sessions were in 2017 (Table 8.1).

To attain the project objectives, COMETIL was in charge of inviting participants to receive explanations on the intervention and schedule the interviews.

8.3.2 Methods of Analysis

First, we described the progression of the intervention and identified the main expansive learning actions performed in sessions. In this description, we emphasized the first and second stimuli used in each session and present concrete outcomes and learning actions.

Table 8.1 Summary of data collection

Procedures	Objectives	What?	With whom?	How long?
Interviews	To understand the activity of each institution To collect mirror data	31 semi-structured individual interviews	28 representatives of public institutions, 2 of trade unions, and 1 of employers	About 16 h (945 min)
Document analysis	To establish TAC and *promo*[a] requests	Two official documents written by MPT		2 h
Introductory workshop	To present the CL method to the group which participated in COMETIL in 2016	Discussion/definition of problem monitoring	Group of professionals: CEPRESOM, municipal secretariats, SENAC, trade unions, universities	About 3 h
CL sessions	To establish COMETIL activity systems to deal with child labor	CL sessions: 5 sessions in November and December 2016 and 2 follow-up sessions in November and December 2017	Professionals from public institutions which participated in COMETIL	14 h (2 h per session)

TAC behavior adjustment agreement, *MPT* Labor Prosecution Office
[a]Operational procedure elaborated by the MPT that aims to accomplish preventive actions related to Brazilian labor laws

8.4 Challenges Posed by the Runaway Object

Based on the collected ethnographic data, we confirmed that COMETIL was not a single activity system (AS); that its object was not transparent, albeit shared; and that the Committee had difficulties to achieve the expected results. We formulated at least three hypotheses on the main problems.

The first concerned the difficulty of some actors to recognize child labor as a social problem. As some of them argued: "In the end, children are doing nothing but to help their family… They're off the streets… They're learning a job… They're not using or dealing drugs."

> CEPROSOM gives us feedback, "Please, notice, a teenager has stopped coming!" Then we go and warn his mother: "Please, notice, your son will remain in the course, because it's a guaranteed right!" We've already participated in other committees and, for example, I've even found school principals who support child labor: "One of the children works at my husband's jewelry factory" … Got it? (CEPROSOM social worker)

The second main problem concerned the difficulty of the production sector to recognize its responsibility (derived from outsourcing):

> There's no child working in the companies we represent. We have no effective proposals, but we look for orientation for companies, and sought to establish whether our industry directly caused the problem. However, to the best of our knowledge, inspections did not

> detect this issue. Employers know the law and seek shelter in the law (Technician representing employers).

The last problem concerned the difficulty of governmental actors to implement actions for promotion of cultural changes:

> Work at home is the most dangerous. So, we tried to sensitize and persuade other institutions to understand its transversal nature, and the need to discuss these issues. We're strategically trying to persuade them, and then develop partnerships between institutions and the training of agents more thoroughly (Professor)

These disturbances might be understood as manifestations of contradictions, such as:

- Low-income families aware of the health risks posed by child labor do not have options for decent jobs.
- The production sector might be pressed by competition with Chinese goods and needs to outsource crucial steps of production to cut costs, increase production, etc., to increase profit.
- Child labor eradication is a secondary object which competes with the primary objects of some public institutions which participate in COMETIL.

These data led us to the hypothesis that COMETIL was a runaway object. Therefore, one of the first objectives of intervention was to facilitate envisioning such object. Next, we describe the trajectory of intervention and its results in terms of learning.

8.5 Learning During CL Sessions

Proper intervention began with a discussion and definition of the problem and of the difficulties of civil society, and mainly the local entrepreneurs, to recognize it.

The aim of the first session was to induce the participants to collectively commit to the need for change. They mapped the problem, which contributed to analysis and to the formulation of new practices. As first stimulus, we used a fable dealing with attempts at solving a complex problem. The leading questions were: Is there any limitation in the activities developed considering eradication of child labor as object? What would such limitations be? Do we need to change? What is COMETIL? What is its purpose? What is the social justification for COMETIL? In which contexts was it created?

The following stand out among the needs listed: preventive actions, educational activities, forums to discuss the interorganizational assistance provision network (from now on simply mentioned as *network*), identification of locations and families where child labor occurs, actions following detection of cases, determination of the responsibility for cases, and most vulnerable neighborhoods requiring urgent actions.

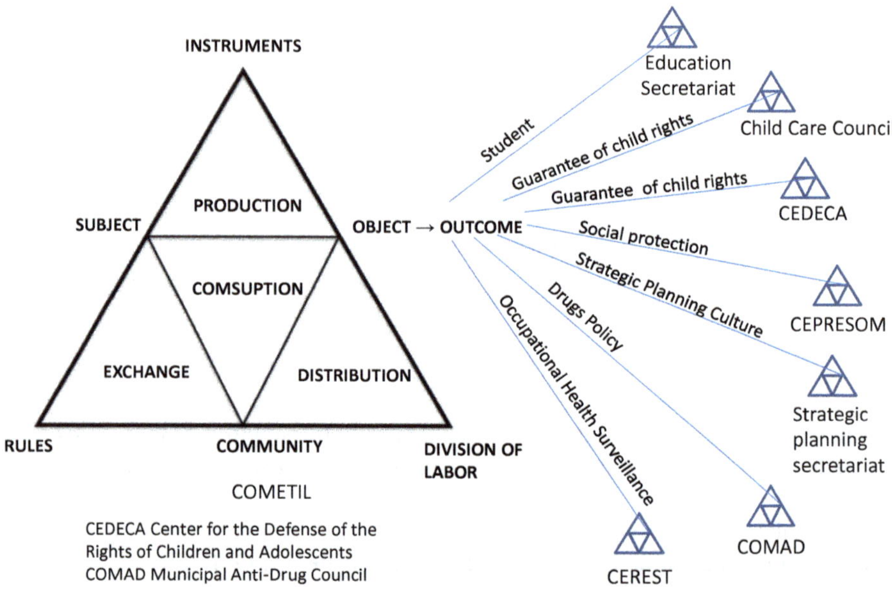

Fig. 8.1 Objects of the institutions participating in COMETIL as defined in the second CL session (the object of each institution appears on the line connecting to the activity system). (Adapted from Engeström 2015)

One issue stood out among the vast list of problems. Upon returning to their institutions, COMETIL agents had difficulty to implement the actions established by the Committee, because the object of child labor elimination/control was competing and secondary at such institutions. This was an uncomfortable situation, given the need to comply with the TAC signed by the city government and MPT in 2009.

The aim of the second session was to investigate the participants' perception of the shared object "eradication of child labor." As a second stimulus, we suggested to first design the AS of the institutions participating in COMETIL and then to present them to discuss the shared object. Figure 8.1 summarizes how each institution saw its particular object.

In the third session, the discussion aiming at defining a shared object continued. As a first stimulus, we suggested reading the report of an accident involving a child aged 5 who had swallowed acid.[1] This case was essential to the group's reflection and also allowed identifying gaps in the network.

Two contradictions were identified:

- Lack of articulation between subject COMETIL and the community (e.g., the Committee combats child labor, but families and entrepreneurs consider it normal).

[1] Companies and families use acid as raw material in cold welding.

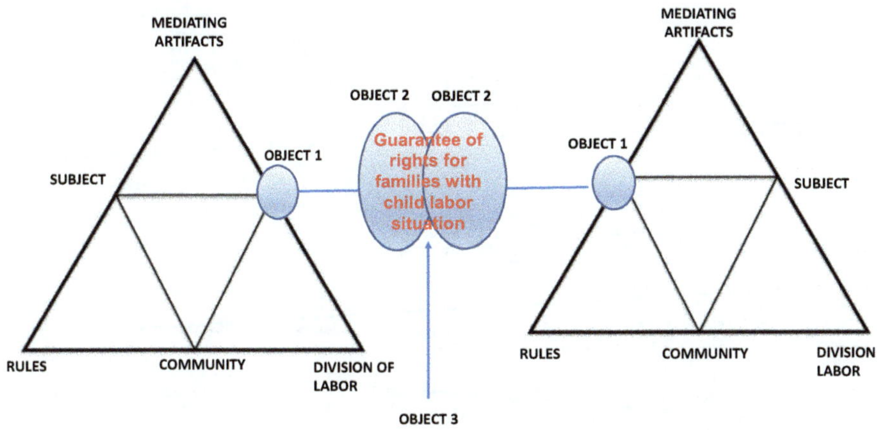

Fig. 8.2 COMETIL shared object as defined in the third CL session. (Adapted from Engeström 2009)

- Lack of articulation between instruments and the division of labor (e.g., the Committee and each institution's role conflicted as a function of hierarchy and lack of a common flow).

The establishment of the various AS, together with the discussion of the aforementioned accident report, led to the first definition of the shared object, which was identified as "guarantee of the rights of families including children who work" (see Fig. 8.2).

In the fourth session, COMETIL object was confirmed, and based on the aforementioned accident report, the participants identified gaps and disturbances in the system. The participants drew a timeline (Fig. 8.3), and with this the aim of the session was accomplished, as it represented relevant facts of COMETIL development and child labor in Limeira. Further analysis of the accident report pointed to some gaps in the healthcare network, including precarious care delivery at primary care services, delay to hospital care, and flaws in the communication among CEPROSOM departments, resulting in delayed Social Assistance Reference Center (CRAS) and Child Care Council (CCC) intervention.

The timeline was highly relevant because it allowed the participants to become aware of several issues: CEREST and MPT collaboration in regard to child labor, changes in COMETIL members from 2013 to 2014, and relevance of GTETIS, which aims are to sensitize the population and change its culture.

In the fifth session, participants designed an AS based on object "Guarantee of the rights of families with children who work" (see Fig. 8.4).

The aim of some COMETIL meetings in 2017 was to implement the network flow established to detect instances of child labor and institutional actions. The definition of AS allowed designing a specific action plan to be tested as pilot project.

The participants also discussed a new municipal decree which restructured COMETIL activities and ensured its existence, which thus became entitled to

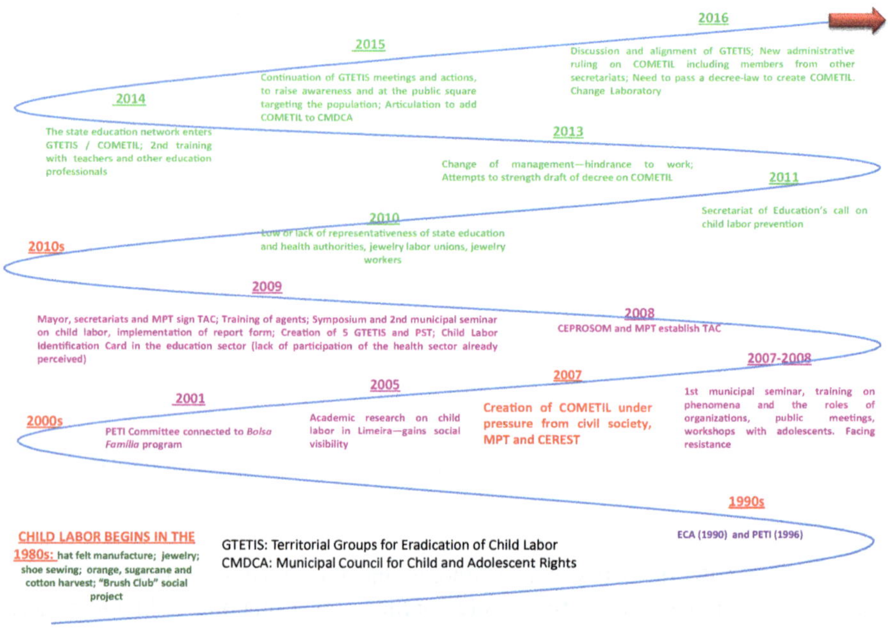

Fig. 8.3 COMETIL timeline as defined in the fourth CL session (*Bolsa Familia* is a federal economic assistance program for incomeless families)

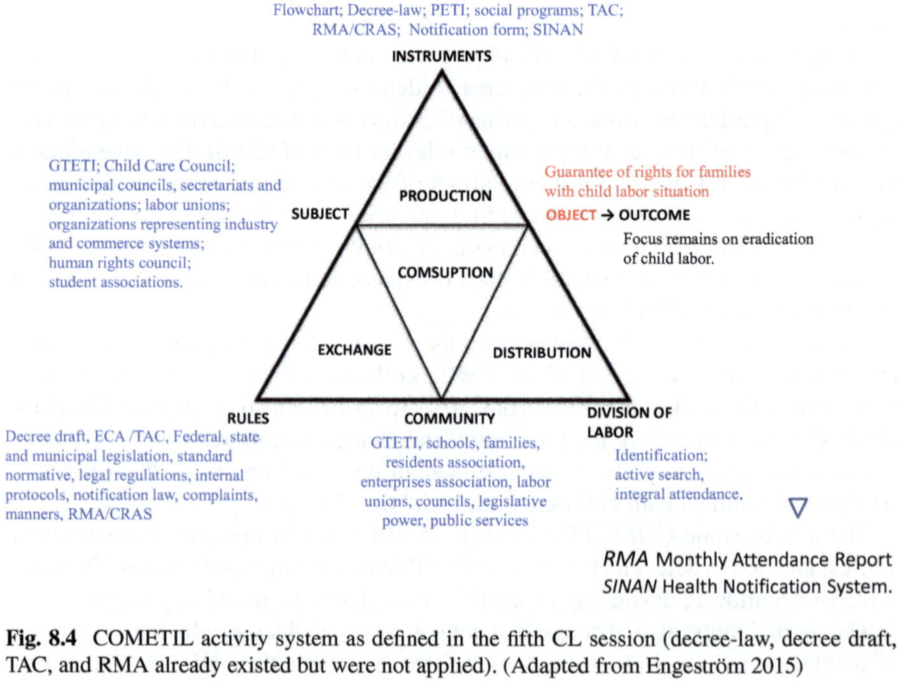

Fig. 8.4 COMETIL activity system as defined in the fifth CL session (decree-law, decree draft, TAC, and RMA already existed but were not applied). (Adapted from Engeström 2015)

"suggest, deliberate on, supervise, and promote actions, projects and policies for prevention and combat of child labor and protection of adolescent work, aligned with the government, municipal councils and civil society in Limeira" (Limeira 2016).

Establishment of the network flow, as instrument, was relevant as a function of the need to keep the focus on:

- Eradication of child labor
- Follow-up and monitoring of specific actions by the municipal institutions which provide social assistance to families
- Prevention of accidents involving children
- Identification of adolescents who work via the Single Registry of Social Programs

To follow up the implementation of the new network flow, the researcher-investigators decided to attend COMETIL 2017 meetings as observers and also to observe a case detected by COMETIL by means of the new network flow.

8.6 Assessment of the New Network Flow Based on a Detected Case of Child Labor

At the end of 2017, we held two follow-up sessions aiming at analyzing the implementation of the new network flow. A case of child labor detected by means of the new instrument was selected as pilot study. The case concerned a family composed of a woman and her 10 children aged 6 months to 21 years old.

School teachers and social workers had noticed that two siblings were always sleepy in class. One day, one of them started scratching his eyes until it caused severe eye irritation, for which reason he was taken to the closest healthcare facility, where he was found to have traces of acid on the hands. Upon being inquired by the physician, the child said he had helped his mother in making jewelry the previous night.

In this particular case, the school was the starting point for detection. As per the new network flow, CRASS was notified, which in turn reported the case to CCC. The institution which first detects a case remains responsible for following it up and must report it to other institutions, in addition to coordinating visits, meetings, and the actions to solve it.

While CCC must be notified and is in charge of enacting the protocols in force, including mobilization of other agents to develop the necessary actions, it is not responsible for all the network procedures. According to the pilot network flow, when the CEPROSOM checks a case, it should report to other professionals included in the network.

In the first follow-up session, we reminded the participants the outcomes of the CL sessions held the previous year, namely, COMETIL's activity object and the new network flow. The participants discussed changes and disturbances along the stage of implementation and who would do what and when.

Several changes were identified along the follow-up of the pilot case study:

- According to the new network flow, the responsibility for monitoring cases was shifted to the institution which first detects them; CCC enters mandatorily in action in compliance with the legislation.
- Priority for implementation of the new network flow at the two neighborhoods with the highest child labor incidence.
- Definition of the three sub-flows, as the hierarchical division of labor at each municipal secretariat made a single flow difficult.
- The new decree granted new roles to COMETIL, which ensure conditions for effective GTETI operation.

A COMETIL agent observed that, as result, GTETI operation improved: "There are many edges, each case is unique, detection is difficult, but we have made advances."

Nevertheless, the main contradiction involving division of labor and application of the new instrument remained. The following dialogue on control mechanisms and case follow-up provides a clue:

- […] CRAS is always notified because it's in the area […]. We work together. Have you detected a case? You'll also call CREAS,[2] won't you?!
- And at school, whom should I call?
- The Child Care Council. You'll register [the case] with the social protection [department] at CEPROSOM headquarters.
- There are things we need to reinforce, we even need to record who's charged of reporting cases.
- Registering and making [the situation] official is one thing, and scheduling a meeting is another. So, it's obvious you'll make the calls and establish the contacts, but you need to register [the case], you have to follow the path to get the data, so you can be in control along follow-up.
- Sometimes you just want to discuss the case. Now, if you want a report with information on the family, you need to register it […] it has to go through the management. Nothing hinders carrying the discussion as we've always done.
- The thing with the protocol is about collecting data for the follow-up of cases. COMETIL and Social Protection don't know whom or which cases are being discussed or how they're evolving […] Most obviously, if the network were strong within the area, the cases would come here, and GTETI would report them, and it would remain responsible for reporting, but this is not the case, therefore, probably someone will end up with an overload of work (dialogue among CRAS social workers).

The aim of the second follow-up session was to verify the development of the COMETIL activity system and to confirm the object defined in 2017 (see Fig. 8.2).

The discussions in the follow-up sessions helped the participants realize that some advances had been made in the approach to detected cases of child labor. Discussions further included preservation and revision of the new flow, monitoring of results, activation of the network before any intervention, collective decision-making on interventions, and formulation of a single report form.

[2] Specialized reference center for social assistance.

Table 8.2 Summary of CL sessions

Session	First stimulus	Second stimulus	Concrete outcomes	Learning actions
1	Are there any limitations to implement practices considering child labor eradication as object? What do we need to change? *Assembleia dos Ratos*[a] to encourage discussions on the need to solve a specific problem	List of problems Diagram of factors which influence work Workflow Action plan	Commitment to change List of daily disturbances in activities	Recognizing and questioning about the problem A problem exists. How and with whom can we work to solve it?
2	For what purpose was COMETIL created? What is its objective? What is its social justification? In which context was it created? What are TAC objectives? How did we work before TAC?	AS model Concept of "object"	Design of the AS of each institution participating in COMETIL	Difficult integration between institutions (each has its own rules, and there are too many of them)
3	In which areas actions were not sufficient or not developed? How might we explain the detected disturbances? Is child labor eradication an expected outcome or an object? Work accident case (child who ingested acid)	Based on the discussion of the work accident case study: AS model Concept of "object" and "contradictions"	List of actions (complete) "Object" definition (partial)	Object definition (variable over time due to the difficulties to solve it)
4	How did child labor emerge in Limeira? How have public policies addressed it historically? Work accident case (child who ingested acid)	To detect the main gaps and disturbances to be solved in a new AS model based on the discussion of the work accident case	Identification of some gaps in the assistance network	Need to cross boundaries: actions are isolated and do not function as a system

(continued)

Table 8.2 (continued)

Session	First stimulus	Second stimulus	Concrete outcomes	Learning actions
5	Discussion of COMETIL current work process flowchart Based on the new systemic view, what might we change in the flowchart?	AS model elements	Design of flow draft to be improved and tested in May 2017	Need for specific flows for each institution in the network
6 (follow-up)	Discussion of the AS established in the 5th session. Did it work as a new model? Evaluation up to the point where we stopped. How do we continue? What do we expect from CL? What do we still need to change? Why did the problems occur? Pilot case study (mother with 10 children)	List of main changes	Identification of main changes during the implementation phase	Difficulty to integrate specific flows Lack of participation of some institutions
7 (follow-up)	Presentation of the new flowchart to detect gaps to be filled in the future Pilot case study (mother with 10 children) Do the other cases mentioned contribute to detection?	Development of AS "Who does what" guideline	Identification of strategies to preserve networking	Need for permanent revision and monitoring of the new flow to detect instances of child labor

[a]Popular Brazilian story based on Aesop's fable "Belling the Cat"

Table 8.2 summarizes double stimulation, results, and learning gained in CL sessions. Learning actions are depicted in Fig. 8.5.

8.7 Discussion

Although the participants reassessed the pilot case study from a collective perspective, they could not solve it. They collectively discussed social guarantees to the involved family. CCC addressed requirements, such as housing, minimum income, keeping the children at school, and daycare for a baby to be born soon.

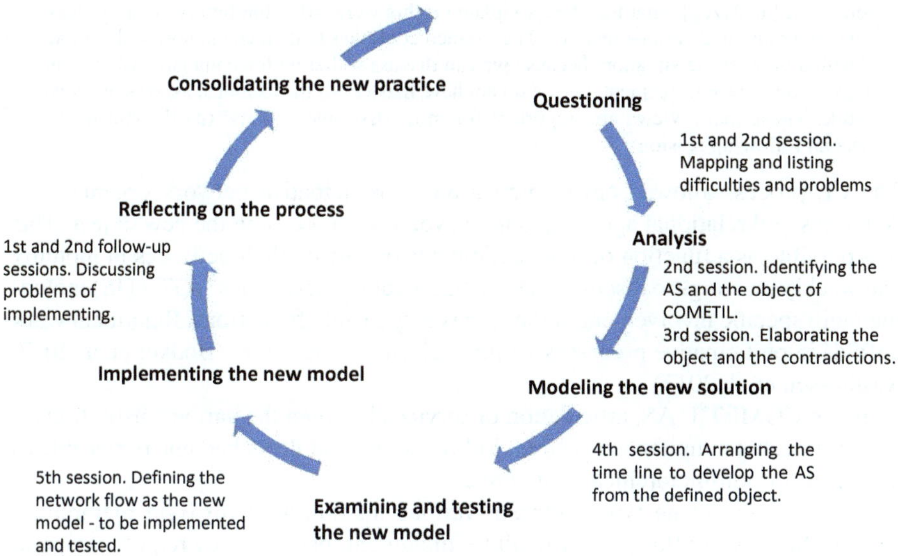

Fig. 8.5 Cycle of expansive learning actions at the COMETIL CL. (Adapted from Virkkunen and Newnham 2013)

Some of the problems which remained in the network were difficulties to overcome resistance vis-à-vis child labor among families, health services and employers. In regard to the pilot case study, the involved woman still makes jewelry, which is her main source of income. She does not see the available public aid as hers and her family's right. As an agent explained: "She believes it's for desperate people! As if it were charity and an insult to her honor!"

Similarly, also others health professionals who participated in COMETIL meetings mentioned having detected some cases of child labor but did not see it as a problem. Another COMETIL agent observed: "In the meeting in which the case was discussed, the health [service staff] said it's no problem at all. Children are off from the streets, and the mother does play her protective and caring role in the family: what else do you want?!"

This type of situation creates barriers at work, weakens the ties of trust between professionals and families. Integrality between health services would allow modifying the approach to those who need healthcare, but there are difficulties to achieve such integration (Favret-Saada 1990).

The changes agreed by the participants consist of exchanges during network meetings and generalized use of the WhatsApp application, understood as an instrument which enables faster communication on cases, even before reports reach their destination.

> The issue of dialogue between network professionals is important because it speeds up [the process]. We understand better that family context and what interventions are needed. You get an earlier view of the whole and its parts, of what each one will do. This is different from how it was before. The Child Care Council first visited the household, checked the

> situation, [analyzed] what the child's violated rights were, what the family situation was, and then requested various services. The Council continues to do this, but now with a more detailed view of the situation, because we can discuss it first with technicians. We master legal issues, law enforcement, but we do not have mastery of the technical aspects of social work! These many views are important for more assertive actions! (participant in the second follow-up session)

The CL process allowed envisioning a more participative network organization. Activities and relational agency began to evolve with focus on the new object. This was possible as a function of how participants organized their activities in monthly and weekly meetings to discuss cases in the so-called "territories" (GTETIS) including more specific involvement of the agents responsible for actions. Relational work unfolds in participative processes within a given infrastructure (Bødker et al. 2017; Rasmussen et al. 2017).

In the COMETIL AS, articulation of services between the various institutions is not organic, but simultaneous and multidirectional, and flows become permeable as is typical for non-hierarchical structures.

Tools to coordinate work between various institutions with rigid hierarchies, such as the network flow, are difficult to implement, because they require cooperation from different institutions and engagement of frontline agents, as is the case of COMETIL professionals. Transformative agency developed among actors inasmuch as their individual actions demanded interacting with other institutions (Peixoto 2014; Bødker et al. 2017). According to Giddens:

> [...] actors not only monitor the flow of their activities continuously and expect others to do the same for their own; they also routinely monitor aspects, social and physical, of the contexts in which they move. (Giddens 1984, p. 5)

The new flow became the instrument to change the pattern of agency. Previously, the approach to an instance of child labor followed CCC legislation. Currently, each case is seen locally, and the CRAS involved becomes responsible for the entire follow-up until the case is closed, as shown by the following dialogue between agents:

- CL is helping COMETIL systematize a networked workflow methodology.
- CL was the tool which helped us go deeper in the discussion on the flow and understand it better within this system of activities (see Fig. 8.5).

Network mobilization allowed the institutions share tasks and afforded the possibility to identify the technicians responsible for monitoring cases. In the past, only the involved agents made personal contact and monitored cases, and the process was delayed due to the need to wait for the delivery of documents, which made referrals even more difficult.

Once they understood their shared object, the participants were able to envision the need for the network to operate in a clear and objective way, and sharing information between municipal secretariats became vital (Star and Ruhleder 1996; Bødker et al. 2017; Rasmussen et al. 2017).

At the time when the network flow first began to be tested, agency developed toward the new object. Cooperative solutions among agents did not stop while they waited for the reports.

WhatsApp application behaved as an innovative instrument; included as means of exchange, it enabled backstage work. It allowed overcoming difficulties associated with the division of labor among the various institutions (Bødker et al. 2017; Rasmussen et al. 2017; Haapasaari et al. 2016). In addition, it afforded new possibilities of operation and thus resulted in expanded horizons. The use of WhatsApp as a communication tool to discuss and accelerate the process of assistance is a small innovation, but its potential for expansion is difficult to measure. The same is the case of the runaway object addressed in the present study, which scope cannot be controlled. Also, side effects are unpredictable, as if they had a life of their own.

We might infer that there are invisible tissues able to grow, survive, and look for escape routes from problems independently from their complexity. These are difficult-to-connect structures, closed or dead, but also vulnerable (Engeström 2006; Bødker et al. 2017).

8.8 Conclusions

Dealing with runaway objects poses an enormous challenge. As a rule, they penetrate structures, and their boundaries are difficult to establish and control. Combating such objects depends on the cooperation of different AS.

Resource to the concept of runaway object evidenced tension between rigid rules (in the legislation for child and adolescent protection), institutional hierarchy, and the need to provide urgent help. Help is urgent, because needs are more dynamic than information and undergo continuous changes which demand immediate actions, which do not depend on a single command center.

The AS established by the participants, as a first attempt at developing a unit of analysis, evidenced the Committee's potential. It also pointed out its limitations, which collide with the hierarchical structures of institutions, given the difficulties to implement the new workflow.

The CL made the object visible and allowed reshaping it and transforming it into an object shared by different AS. Intervention moved the object to a dimension closer to reality, as the participants began acting in a more independent manner within the rigid structures, thus creating possibilities to aid families with children and adolescents who work.

The CL method proves to be a tool that enabl analyses of historical relations within invisible, hard-to-deal-with social tissues, which are consequently marginalized because they do not yield patent profit. Interventionist methodologies allow for systemic understanding of events and thus afford expansive learning to the involved actors (Engeström 1987, 2015) (see Fig. 8.5).

In traditional research methods, participants wait for solutions from experts, while in CL collective construction is expected. Hence, studies involving runaway objects, generally of a complex nature, might benefit from this method, because it affords a systemic view, leaving aside recommendations made by experts. Collective formulations of solutions for complex problems enable engagement in concrete actions.

References

Bødker, S., Dindler, C., & Iversen, O. S. (2017). Tying knots: Participatory infrastructuring at work. *Computer Supported Cooperative Work, 26*(1–2), 245–273. https://doi.org/10.1007/s10606-017-9268-y.

Engeström, Y. (1987). *Learning by expanding: An activity-theoretical approach to developmental research*. Helsinki: Orienta-Konsultit.

Engeström, Y. (2006). Development, movement and agency: Breaking away into mycorrhizae activities. In K. Yamazumi (Ed.), *Building activity theory in practice: Toward the next generation* (pp. 1–43). Osaka: Center for Human Activity Theory, Kansai University.

Engeström, Y. (2009). The future of activity theory: A rough draft. In A. Sannino, H. Daniels, & K. Gutierrez (Eds.), *Learning and expanding with activity theory* (pp. 303–328). New York: Cambridge University Press.

Engeström, Y. (2015). *Learning by expanding: An activity-theoretical approach to developmental research*. New York: Cambridge Press.

Engeström, Y., Virkkunen, J., Helle, M., Pihlaja, J., & Poikela, R. (1996). The change laboratory as a tool for transforming work. *Lifelong Learning in Europe, 1*(2), 10–17.

Favret-Saada, J. (1990). Être affecté. *Gradhiva: revue d'histoire et d'archives de l'anthropologie, 8*, 3–9.

Ferreira, M. A. L. (2005). *Estudo de riscos à saúde do trabalhador e ao meio ambiente na produção de joias e bijuterias de Limeira–SP* [dissertation]. Piracicaba: UNIMEP, Portuguese.

Giddens, A. (1984). *The constitution of society: Outline of the theory of structuration*. Cambridge: Polity Press.

Haapasaari, A., Engeström, Y., & Kerosuo, H. (2016). The emergence of learners' transformative agency in a Change Laboratory intervention. *Journal of Education and Work, 29*(2), 232–262. https://doi.org/10.1080/13639080.2014.900168.

Lacorte, L. E. C. (2012). *A construção de políticas públicas em rede intersetorial para a erradicação do trabalho infantil em Limeira-SP* [dissertation]. São Paulo: Universidade de São Paulo, Portuguese. doi:https://doi.org/10.11606/D.6.2012.tde-10092012-162500.

Lacorte, L. E. C., Vilela, R. A. G., Silva, R. C., Chiesa, A. M., Tulio, E. S., Franco, R. R., et al. (2013). Os nós da rede para erradicação do trabalho infanto-juvenil na produção de joias e bijuterias em Limeira – SP. *Revista Brasileira de Saúde Ocupacional, 38*(128), 199–215. https://doi.org/10.1590/S0303-76572013000200009.

Limeira, (2016). Decreto n° 399 de 22 de dezembro de 2016. Institui a comissão municipal de erradicação do trabalho infantil e proteção do adolescente no município de Limeira. Jornal Oficial do Município. 28 Dez 2016.

Peixoto, M. A. (2014). Estrutura e agência em Anthony Giddens: uma análise crítica do estruturacionismo. *Revista Sociologia em Rede, 4*(4), 93–106.

Rasmussen, M. B., Gade, A. N., & Rasmun Lund, J. (2017). A knotworking guideline for building projects: Bridging the gap between participants. *International Journal of Engineering Research and Applications, 7*(6 (Part 4)), 24–40. https://doi.org/10.9790/9622-0706042440.

Star, S. L., & Ruhleder, K. (1996). Steps toward an ecology of infrastructure: Design and access for large information spaces. *Information Systems Research, 7*(1), 111–134. https://doi.org/10.1287/isre.7.1.111.

Vilela, R. A. G., & Ferreira, M. A. L. (2008). Nem tudo brilha na produção de joias de Limeira – SP. *Production, 18*(1), 183–194. https://doi.org/10.1590/S0103-65132008000100014.

Virkkunen, J., & Newnham, D. S. (2013). *The Change Laboratory: A tool for collaborative development of work and education*. Rotterdam: Sense Publishers.

Chapter 9
Challenges to Change Laboratory Learning in a Dynamic and Complex Civil Construction Project

Manoela Gomes Reis Lopes, Rodolfo Andrade de Gouveia Vilela, Marco Antonio Pereira Querol, and Ildeberto Muniz de Almeida

Abstract The chapter presents a Change Laboratory (CL) intervention in the construction airport activity. After the occurrence of two severe accidents in a brief time interval, the Labor Prosecution Office sought the School of Public Health to assist at the second accident analysis. The application of Analysis and Prevention of Accidents' Model (MAPA) and CL appeared as a proposal. During the implementation process, the activity was in a crisis phase. The participants identified the problems, but they felt paralyzed in some moments. The CL allowed the participants to formulate hypotheses of the main problems and their relation with historical changes in the activity. The CL allowed the creation of solutions, but with the shutdown of the construction site, it was not possible to have continuity in the implementation of solutions. In the chapter will be presented the results achieved, the difficulties faced, and proposition of suggestions for actions to improve and continue the development of the activity.

M. G. R. Lopes (✉)
Department of Biological Sciences and Health, Federal University of Amapá, Macapá, AP, Brazil
e-mail: lopes.manoela@unifap.br

R. A. G. Vilela
Department of Environmental Health, School of Public Health, University of São Paulo, São Paulo, SP, Brazil

M. A. P. Querol
Department of Agronomic Engineering, Federal University of Sergipe, São Cristóvão, SE, Brazil

I. M. de Almeida
Department of Public Health, Botucatu School of Medicine, São Paulo State University "Júlio de Mesquita Filho", Botucatu, SP, Brazil

R. A. G. Vilela et al. (eds.), *Collaborative Development for the Prevention of Occupational Accidents and Diseases*, https://doi.org/10.1007/978-3-030-24420-0_9

9.1 Introduction

Civil construction in Brazil received incentive in recent years to build infrastructure by means of the Growth Acceleration Program (Programa de Aceleração do Crescimento—PAC) (Brasil 2013). Along this time, the country also hosted the FIFA World Cup and the Olympic Games, with consequent increase of the demand to build stadiums, airports, and highways. The civil construction sector exhibits high rates of work accidents, being characterized by multiple hazards, high turnover rates, precarious working conditions, occupational diseases, and work accidents (Costa 2013; Véras 2014).

Within this context, work accidents are considered a serious public health problem, whence prevention is relevant. For this purpose, accurate knowledge of the multiple proximal and latent factor network involved in the origins of these events is necessary.

Investigation of accidents still commonly involves analyses which attribute events to a few causes; the usual conclusions are unsafe acts or conditions, whereby workers are held responsible and guilty. The approaches centered on monocausal analysis have been replaced by systemic views in recent years.

One offshoot of this shift is the development of methods for analysis, such as the Analysis and Prevention of Accidents' Model (Modelo de Análise e Prevenção de Acidentes do Trabalho—MAPA) (Almeida and Vilela 2010; Almeida et al. 2014) and Event Organizational Analysis (EOA) (Llory and Montmayeul 2010; Dien et al. 2012).

However, these new in-depth methods for accident analysis still have to deal with some difficulties, particularly as concerns the implementation of recommendations for organizational changes.

It is believed that this is partly due to limitations to the involvement of workers, and even also of managers, in analysis, as well as to the fact that when they do participate, it is a mere informant. From a practical perspective, one might say that analysis is monopolized by analysts/experts and that the participation of decision-makers tends to occur only at the end of intervention, i.e., after decisions have already been made on proposals considered likely to improve the situation. In addition, analyses of accidents are not grounded on learning theories, which impairs organizational learning.

Cultural-Historical Activity Theory (CHAT) and the Change Laboratory (CL) (Engeström et al. 1996; Engeström 2007; Virkkunen and Newnham 2013) have the potential to overcome the aforementioned limitations and contribute to the design of a new type of intervention on sociotechnical systems for analysis and adoption of accident prevention measures. The studies in this regard are still scarce. The aim of the present chapter is to describe the main advances and difficulties detected during the performance of a CL for work accident prevention in civil construction sector.

This is an innovation in health and safety, as well as one of the earliest examples of application of CL in this field in Brazil and worldwide. Therefore, in this chapter, we intend to answer the following questions: What were the main advances and difficulties found during the performance of CL? What strategies and learning are needed to overcome such difficulties?

9.2 Study Context

To answer the research questions, we had resource to an empirical case, to wit, the airport reform and expansion of an airport. Two cases of work accidents occurred along this process over a short period of time. One involved an earthfall where 1 worker was buried and died; on the other, the concrete slab fell and hurt 16 workers. The latter took place at night, while a beam was being filled with concrete; the structure collapsed, and 16 workers fell from a height of about 10 m together with it.

After the last accident, the Labor Prosecution Office (MPT) sought the School of Public Health, University of São Paulo (FSP/USP), to collaborate in the analysis of the second accident.

After a bid for a 30-year concession, the winning company (concessionaire) became responsible for the management and also expansion of the airport infrastructure by building a new passenger terminal, runways and aprons, a parking lot, and boarding and disembarking piers. To perform this work, the company hired a contractor (a joint venture).

In the airport construction activity system (see Fig. 9.1), the joint venture was considered as the unit of analysis and thus as the subject, in which object was infrastructure construction, the division of labor involving outsourced firms and company departments such as the Health, Safety, Environment, and Social Responsibility (HSE-SR), engineering, planning, quality, procurement (responsible for the acquisition of supplies), human resources, and production departments. The instruments used included schedules and contracts with deadlines and delay penalties. Rules included technical standards, concession, safety, and labor rules. The joint venture interacted with a community which included public regulatory organs, such as MPT, the Ministry of Labor and Employment, a Workers' Health Reference Center (Centro de Referência em Saúde do Trabalhador—CEREST), the National Civil Aviation Agency (Agência Nacional de Aviação Civil—ANAC), and the concessionaire (Lopes et al. 2018a).

The construction work was influenced by a political agenda which did not take the time it actually required into account. According to the workers, the deadline set for the construction to be concluded was, as a fact, the time needed to design the executive project, which resulted in contradiction between a complex object and unrealistic deadlines (see Fig. 9.1a). In addition, the joint venture had been hired under the Engineering, Procurement, and Construction (EPC) regimen, which made it responsible for all the stages of work: acquisition of materials, project design, and construction. Yet this company had no expertise in project design, which in fact contributed to the contradiction detected between the subject and object (see Fig. 9.1b). Given the time pressure, this contracting arrangement allowed developing the various stages of the project construction concomitantly by outsourcing,

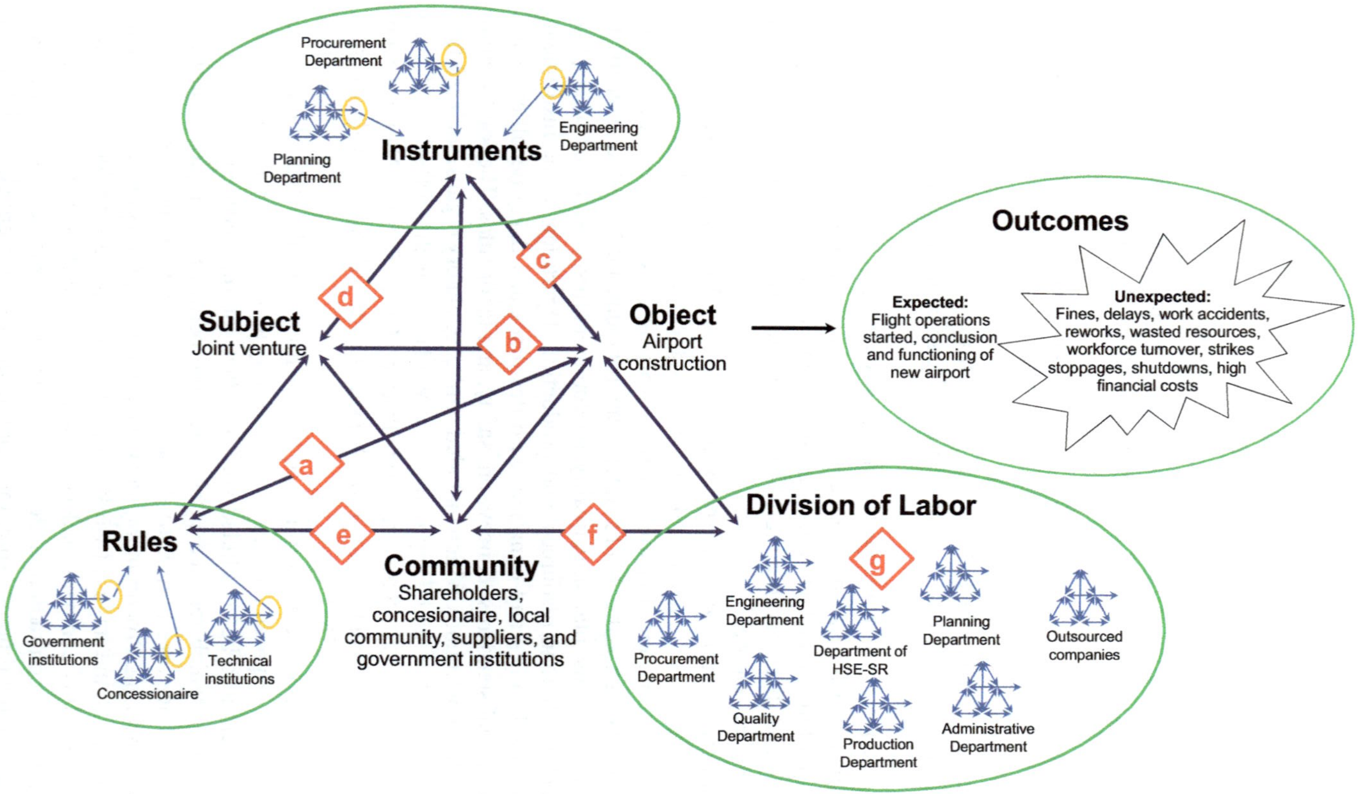

Fig. 9.1 Airport construction activity system and its contradictions. (**a**) Contradiction: rules *vs.* object. (**b**) Contradiction: subject *vs.* object. (**c**) Contradiction: instruments *vs.* division of labor. (**d**) Contradiction: rules *vs.* instruments. (**e**) Contradiction: rules *vs.* division of labor. (**f**) Contradiction: community *vs.* division of labor. (**g**) Contradiction between elements in the division of labor. (Adapted from Lopes et al. 2018a)

however, without due coordination among the various outsourced firms and the joint venture departments. The result was contradiction between the nonstructured schedule (instrument) and division of labor, with overlapping of activities (see Fig. 9.1c). The joint venture had formulated some internal regulations aiming at developing an organizational culture of its own. However, this process requires time, while the actors were under time pressure to meet deadlines. The result was contradiction between the joint venture's internal operational rules and other components of the activity system, such as the object, instruments, and division of labor (Lopes et al. 2018a) (see Fig. 9.1a, d, e).

Another relevant aspect was the high turnover rate of directors. To be sure, four different construction directors were appointed in the course of 2 years, each of whom changed the organizational structure and managers, resulting in contradiction in the division of labor (see Fig. 9.1g). These events led to 1-month delays, on average to each director change, for workers to adapt to the new procedures. In addition, by that time, Brazil was undergoing a construction boom, which made hiring the best professionals difficult and led to contradiction between the community and division of labor (Lopes et al. 2018a).

All these factors increased the need for reworking and resulted in waste of financial resources, with the consequent negative impact on the schedule and increase of the time pressure. The effects were disturbances, such as high turnover rates, occupational diseases, and work accidents (Lopes et al. 2018a).

9.3 Methodological Trajectory

Here we describe the qualitative approach to the analysis of an accident which took place during the airport expansion work. Data collection was performed in two stages: ethnographic and CL sessions.

Ethnographic data collection was performed along 54 days of fieldwork, to a total of 378 h along 9 months. In this stage we analyzed 92 documents, observed activities, and performed 132 individual or collective interviews with employees from several joint venture departments and allocated to different functions, as well as with employees of the outsourced companies. Two Collective Work Analysis (Análise Coletiva do Trabalho—ACT) sessions (Ferreira 1993) were also performed with 29 employees with jobs somehow related to the second accident, to a total of 4 h.

For ACT, voluntary groups of workers are formed to talk about their job activities, including the participation of at least two researchers. Meetings should be held away from the workplace and last about 90 min. The orienting question is "What do you do at your job?," the answers to which are thoroughly discussed in sessions (Ferreira 1993).

Ethnographic data collection included observation of work situations, individual and collective interviews, document analysis, accident analysis, and so forth. This stage provided the empirical data needed to analyze the accident of interest based on

MAPA (see Chap. 14) and EOA (Llory and Montmayeul 2010; Dien et al. 2012). These information also served as mirror data in the CL sessions (Engeström et al. 1996; Engeström 2007; Virkkunen and Newnham 2013).

Following negotiation with the joint venture management and legal department, the number and length of sessions and the criteria for participation in CL could be established. A total of 6 weekly CL sessions were held, lasting 2–3 h, to a total of 15 h, with 11 employees, on average.

The CL sessions were planned based on the double-stimulation method and the expansive learning cycle (Engeström 1987, 2007, 2015) to stimulate participants' agency and expansive learning.

During CL sessions, the investigators sought to stimulate the following learning actions, in compliance with the method recommendations: questioning of critical aspects and ongoing practices within the activity system; analysis of the current situation and its historical antecedents; identification and analysis of contradictions; visualization and formulation of a new activity system model; formulation of possible solutions; and implementation, testing, and evaluation of the new model (Engeström 1987, 2015).

Interviews, ACT, and CL sessions were recorded and transcribed. Due to ethical reasons, the interviewees' names and involved companies are not indicated.

The study was approved by the research ethics committee of School of Public Health, University of São Paulo, CAAE 11886113.5.0000.5421. Participants' autonomy and anonymity were ensured, as well as the privacy of confidential data, as required by the National Health Council Resolution n. 196/96.

The main procedures performed along the study are summarized in Table 9.1.

9.4 Analysis and Creation of Solutions Along Change Laboratory Sessions

The CL sessions promoted collective discussions among the participants and researchers. The latter provided stimuli to the debate, and based on conceptual models (Engeström 2007), the participants analyzed the main problems seeking to explain their origin as manifestations of contradictions within interactions along the historical development of the system components and/or interactions between them and/or between them and the environment (see Fig. 9.1).

The sessions enabled in-depth discussions on and analysis of the activity system. It is worth observing that Brazilian workers tend to be afraid of discussing their job freely for fear of retaliation (Jackson Filho et al. 2013). For this reason, the sessions were planned to occur in a way that participants could not be identified.

Following the second stimulus, the group met, discussed, and noted down their answers on a sheet of paper that was then posted on a wall. When in the course of sessions, the participants became more confident and acquired agency, and they suggested changes to the activity pattern. The group discussed answers openly all

Table 9.1 Summary of the main procedures performed along the study

Procedures	Objectives	With what and whom?	How long?
Individual and group semi-structured interviews	To understand the usual work routine To understand the history and transverse and horizontal dimensions of the organization To understand changes and barriers related to the concrete slab accident To collect mirror data for CL sessions	132 employees in individual and group interviews: Joint venture employees allocated to different functions and departments Employees of outsourced companies	While duration was variable, most interviews lasted 1–2 h 66 h of taped interviews
Collective Work Analysis (ACT)	To understand the usual work related to beam assembling To collect mirror data for CL sessions	29 participants: Carpenters Assemblers Masons Hod carriers	Each session lasted about 2 h 4 h in total
Situational observation	To understand the usual work routine To understand the construction dynamics To collect mirror data for CL sessions	Employees of the joint venture and outsourced companies	9-month fieldwork
Document analysis	To identify the accident victims and involved companies To learn about the prescribed work To understand the health and safety management policy adopted by the company To understand the technical causes of the accident To identify accident barriers To understand the joint venture history and transverse and horizontal dimensions To collect mirror data for CL sessions	92 documents from various institutions and departments	9-month fieldwork
CL sessions	To identify anomalies and contradictions To achieve an historical understanding of the emergence of contradictions To promote actors' agency and learning To create solutions	16 participants from key joint venture departments	2-month fieldwork 15 h of CL sessions

together, without any concern with anonymity. This newly gained confidence was a positive outcome which enriched the debate and allowed for better use of the time allotted to sessions.

We believe that extensive ethnographic data collection and the engagement of the principal researcher during the ethnographic stage contributed to enrich the debate as a function of an adequate choice of mirror data. On many occasions during data collection, the workers came to see the principal researcher as one of them, which led to the development of bonding and trust (see Table 9.2).

Another aspect that contributed to the debates was the receptivity and support of the health, safety, environment, and social responsibility department at various times along the study. The department granted access to documents and field access to perform interviews and situational observation and also made suggestions on how to conduct the CL sessions. This reinforced the idea to form an internal follow-up group to improve interventions.

The planning of CL sessions complied with the expansive learning cycle (see Fig. 2.4). The participants exhibited agency all along the cycle, which became evident in the historical analysis of problems and contradictions and in the creation of solutions. In their analyses, the participants identified the main strong and weak points of construction work, designed a timeline with the main critical events related

Table 9.2 Actors' learning and roles along the various stages of the study

	Study stages		
Actors	Ethnographic	CL sessions	Feedback
Researchers	Formulation of hypotheses on the accident causes and construction operation Development of trust relationships with operators and intermediating managers Relevance of HSE-SR team internal support Need to form a follow-up group including workers and researchers	How to prepare mirror data How to plan double-stimulation method Detection of participants' difficulties with more complex concepts (activity system, contradictions) Relevance of HSE-SR team internal support Need to form a follow-up group including workers and researchers	How to plan feedback considering the setting and double-stimulation strategies
Participants	Information provision Friendly and trust-based relationship with investigators Limited learning	Understanding of the systemic and historical origins of disturbances and contradictions (accidents, delays, reworking, etc.) Formulation of possible solutions	Limited learning and involvement as observers
Managers	Isolated from the process	Limited learning, because they did not follow up the process	Limited learning, defensive attitude, and denial of the study results

to the construction, designed a vicious circle and the airport construction activity system, and identified contradictions and possible solutions for prevention of undesirable events, such as work accidents and organizational anomalies (Lopes et al. 2018b).

As shown in section "Study Context," by the time the CL sessions started, the airport construction activity system was facing countless problems and undergoing a crisis (double bind). The participants were able to identify the problems quickly but felt paralyzed on some occasions, for instance, during the analysis of contradictions and at the time of creating solutions.

During the step of creating solutions, a corruption scheme involving major construction firms—including the concessionaire—was revealed in Brazil. As a result, there were cuts in the financial resources, and many workers were laid off, which made the problems of the construction project even worse. Some of the workers who had participated in sessions were laid off, and the others were left in a state of uncertainty as to the future of their jobs.

Within such atmosphere of uncertainty and instability, the participants suggested solutions related to planning and local politics; however, they were unsure of whether such solutions could come or not to be actually implemented.

One of the objectives of the demand for the present study was to perform participatory analysis of the accident that had taken place during beam filling based on CHAT and the CL method. Mirror data were provided during the CL sessions, such as the Institute for Technological Research (Instituto de Pesquisas Tecnológicas—IPT) report on the most immediate causes of the accident analysis. Civil construction is characterized by highly dynamic activities and high turnover of workers. At the time of the accident, intensive concreting was taking place, while the CL sessions started about 1.5 year later this activity was very much reduced. Therefore, as a function of the high turnover rate, many of the participants had started working in the construction project only after the accident. This fact hindered the analysis of immediate causes, which was mostly hypothetical and based on the previous experiences of the participants. In addition, the solutions suggested to avoid accidents during concreting could not be implemented in that particular construction project, since as a function of the civil construction dynamics, this activity was no longer performed.

9.5 Difficulties and Challenges to Implementation

The next phase of the expansive learning cycle consists in the implementation of the proposed new activity model. In the case of the airport construction CL, it was difficult to implement the suggested solutions, among other reasons, due to the large number of layoffs and financial cuts, which eventually stopped the construction for a while.

During analysis, the participants discussed organizational determinants which influenced the problems met along the project, including external factors and political decisions. As Fig. 9.1 shows, government institutions were a part of the activity system community and established some rules, as e.g., for the concession and funding of the construction project.

Such community influence became more evident when the corruption scheme involving the concessionaire was revealed. The division of labor was drastically reduced, to the point the construction work was stopped for a while, which hindered the implementation of the suggested solutions.

The negotiation process and demand also interfered with the stage of implementation. The demand for the present study was not made by the company managers but by the researchers and the MPT. The company health, safety, environment, and social responsibility department recognized the relevance of the present study, but it was not clear whether also the higher management levels also acknowledged its significance. At no time along the study, we had access to the company board of directors, but only after the end of sessions. This situation was made even worse by a series of changes in the company management.

Contact with the high management levels took place at MPT when we reported the study results. The company management did not participate in any of the steps of ethnographic data collection, sessions, or planning. As a result, it remained isolated and failed to acknowledge the problems and solutions pointed out by the employees in sessions (see Table 9.2). In turn, at the time of feedback, workers who had not participated in sessions, but had the same hierarchical level as the participants, had a private conversation with the researchers, in which they confirmed the results obtained. At this point, the differences in views and perspectives of the high-level management and legal department vis-à-vis the intermediate-level employees became fully clear. This finding evidences the relevance of multi-voicedness in the performance of CL. In the present case, the voice of the high-level management was not listened to, and these actors could not be engaged in the process.

We believe that the fact feedback took place at MPT might have contributed to the lack of recognition by the high-level management. Then, active participants in CL had the role of mere observers in the feedback stage (see Table 9.2). Even when we emphasized that the study was an instrument aiming at the prevention of new accidents, the legal defense approach is still strong among companies, while MPT is entitled to inspect and punish them, which might have given rise to fear of recognizing some of the research results in the presence of prosecutors.

One further relevant point concerning the difficulties to implement solutions is related to the characteristics of civil construction. Construction is a highly dynamic and ephemerous activity, which involves several stages: earthworks, concreting, electrical installations, and finishing. Once these stages are accomplished and construction is over, the activity motive fully disappears. Many of the suggested solutions went beyond the scope of the involved company and could take long to be implemented.

For the particular case of civil construction, it would be interesting to think of solutions for projects in general, rather than for particular ones. In the interviews, the workers observed that some of the problems exhibited by that particular project occurred also in others and that they tend to be rather old. These observations reinforce the relevance of designing an activity system for major construction works aiming at the prevention of accidents.

The second accident might be explained as the result of the interaction of multiple flaws: in the management of the project as such, outsourced companies, materials, and safety, within a context characterized by increasing pressure for productivity, as shown above.

Given the recurrence of the same problems, we believe that the latent causes of accidents and the activity system contradictions remain untouched, which fact creates common problems for various construction works.

Another difficulty met in the implementation of solutions is related to limitations to the continuity of the activity system expansive and permanent learning. The participants acquired expansive learning along sessions; however, learning needs to be permanent, and, in the present case, it was discontinued too soon (Lopes et al. 2018b) due to layoffs and the interruption of work.

We believe that, in the present case, learning was acquired by the researchers and the participants in the CL sessions, but not by the organization, as a function of the aforementioned difficulties in the stages of formulating and implementing the expansive learning cycle solutions.

9.6 Final Considerations

In the present study, we met many difficulties, such as lack of recognition of problems by the high-level management, within a context of organizational instability, and the fact that decision-making depended on other institutions, which contributed to hinder the implementation of solutions.

Yet, the study also had positive aspects. The engagement and immersion of the main researcher during the stage of ethnographic data collection contributed to the formulation of hypotheses, choice of mirror data, and development of bonding and trust among participants. In this study, this involvement needs to be highlighted. She was in the fieldwork for about 1 year, in close proximity, especially with members of HSE-SR, creating a strong relationship of trust and bonding. During the research, she came to be seen as someone from the company. We think it was an important strategy to deal with so many adversities during the ethnographic data collection and CL sessions and for engaging workers during the intervention.

Moreover, the safety department kept the same management team even with all the changes of directors. This department supported the research team all the time and we had a very good interlocution with them to gain access to the employees, documents, and everything necessary to the research. This kind of support is not common, mainly because of the legal approach.

The initial demand, together with the legal approach to accidents, did not promote an environment favorable to research. Amidst all these adversities, we were able to establish a rich participatory diagnosis. However, difficult access to the high-level management and the dynamics and high turnover rates characteristic of civil construction contributed to hinder the attempts at implementing the suggested recommendations.

The many changes in high-level management and the large number of outsourced companies—particularly for the manufacture and transport of the components necessary to construction—allow emphasizing the relevance of the presence of representatives of the system higher hierarchical levels in the negotiation of the intervention process, formation of the group charged of data collection and analysis, and of the changes to be selected and implemented aiming at accident prevention.

A political conjuncture in Brazil, with delays in the schedule of construction works associated to events of global visibility, was the source of time pressure and acceleration of the pace of activities. At the same time, problems also occurred in the funds transfer schedule, resulting in additional management difficulties, besides the adoption of a contract management system which demanded from the joint venture to deal with activities for which it had no expertise. Moreover, the simultaneous construction boom made hiring specialized workforce difficult, for which reason workers with little or no experience at all were hired. Finally, as the airport construction was close to its end, the discussion of the final recommendations led to a commitment of implementing them in future projects but not at that specific company.

After data analysis, we believe it is necessary to create a nationwide agenda to review the management model for major construction works, rather than to act upon isolated projects of single companies The CL sessions helped and had potential to implement this agenda, but unfortunately external pressures hindered this process.

One of the main problems evidenced in the present study was the place allocated to safety in the strategic management of the project. Workplace safety had a secondary place and with little autonomy by comparison to other decision-makers. Shortly, the detected situation indicates that the path will be long before the safety agenda becomes one of the integrated and strategic management aspects of this type of project in Brazil. Safety management should be a top priority to be integrated into the strategic agenda since the beginning, as it was seen during the London 2012 Olympics Park construction. This construction had no fatal accidents, being delivered on time and within the budget (Gibb et al. 2018).

During sessions the participants exhibited difficulty to assimilate CHAT and CL concepts. In order to tackle this situation, in today studies we are including training workshops on methods during the stage of ethnographic data collection. Workshops of this type were conducted in other cases of a railway company, CASA Foundation, and a hospital hemodialysis unit (see Chaps. 4, 5, 7, and 15).

The CL-grounded intervention research study conducted within the context of the airport reform and expansion was one of the first experiences of our research group with this method and its supporting theories and concepts. The process described here sought to emphasize learning opportunities and detecting the participants' limitations. Even greater emphasis was given to the need to understand and develop a process which has simultaneous technical, social, and political dimensions.

In this early stage, the advances seem to preferentially concern the first dimension, i.e., the appropriation of notions for application in real situations. In regard to the social dimension, this experience showed that the involvement of researchers in fieldwork and dialogue with more than 100 interviewees was not only feasible but allowed "reconstructing" a major accident and explaining it in an innovative manner, including the participation of internal actors of the activity system.

Nevertheless, more significant advances seem to require creating moments for dialogue and negotiation with the high-level management, this aspect being understood as a goal to be pursued from the very beginning of the process. This was not the case in the present study intervention, partly because the researchers did not seek to create such moment in the earliest stages of the study. The present study evidenced that intervention was performed within a socially and politically complex system and, worse, under unfavorable political circumstances.

It is important to mention the political difficulties faced during this research. During a short period of time, there were changes in the government's contracting of large construction works, followed by delays in scheduling, difficulties in obtaining environmental licenses, and multiple changes in the company's high management; besides, a major corruption scheme involving members of the company's top leadership was revealed.

In addition, delays in the initial phase of the work schedule gave rise to strong time pressures during all the stages, since the facilities were planned to be in use during the FIFA World Cup. For all this, it was not possible to use the recommendations of the new activity model designed during the CL sessions for this construction site.

Finally, it is worth mentioning that this study showed evidences that go beyond identifying problems in a construction site. In our opinion, the study points to the collapse of Brazilian model of large work management. This process could benefit from the use of methods such as CL.

This study was funded by the National Council of Scientific and Technological Development (CNPq), the Brazilian Federal Agency for Support and Evaluation of Graduate Education (CAPES), and the Santander Group.

References

Almeida, I. M., & Vilela, R. A. G. (2010). *Modelo de análise e prevenção de acidente de trabalho – MAPA*. Piracicaba, Brazil: CEREST.

Almeida, I. M., Vilela, R. A. G., Silva, A. J. N., & Beltran, S. L. (2014). Modelo de Análise e Prevenção de Acidentes – MAPA: Ferramenta para a vigilância em Saúde do trabalhador. *Ciência & Saúde Coletiva, 19*(12), 4679–4688. https://doi.org/10.1590/1413-812320141912.12982014.

Brasil. Ministério do Planejamento. Sobre o PAC. 2013. http://www.pac.gov.br/sobre-o-pac. Acessed 4 Jun 2018.

Costa, L. R. (2013). *Trabalhadores em construção: Mercado de trabalho, redes sociais e qualificações na construção civil*. Curitiba, Brazil: Editora CRV.

Dien, Y., Dechy, N., & Guillaume, E. (2012). Accident investigation: From searching direct causes to finding in-depth causes – problem of analysis or/and of analyst? *Safety Science, 50*(6), 1398–1407. https://doi.org/10.1016/j.ssci.2011.12.010.

Engeström, Y. (1987). *Learning by expanding: An activity-theoretical approach to developmental research*. Helsinki, Finland: Orienta-Konsultit.

Engeström, Y. (2007). In H. Daniels, M. Cole, & J. M. Wertsch (Eds.), *Putting Vygotsky to work: The change laboratory as an application of double stimulation* (pp. 363–425). Cambridge: Cambridge University Press.

Engeström, Y. (2015). *Learning by expanding: An activity-theoretical approach to developmental research*. New York: Cambridge University Press.

Engeström, Y., Virkkunen, J., Helle, M., Pihlaja, J., & Poikela, R. (1996). The change laboratory as a tool for transforming work. *Lifelong Learning in European, 1*(2), 10–17.

Ferreira, L. L. (1993). Análise Coletiva do Trabalho. *Revista Brasilera de Saúde Ocupacional, 78*(21), 7–19.

Gibb, A., Lopes, M. G. R., Vilela, R. A. G., & Almeida, I. M. (2018). The successful safety and health experience in the London 2012 Olympic Park construction: An interview with Alistair Gibb. *Revista Brasileira de Saúde Ocupacional, 43*(supl 1), 1–9. https://doi.org/10.1590/2317-6369000020218.

Jackson Filho, J. M., Vilela, R. A. G., Garcia, E. G., & Almeida, I. M. (2013). Sobre a "aceitabilidade social" dos acidentes do trabalho e o inaceitável conceito de ato inseguro. *Revista Brasileira de Saúde Ocupacional, 38*(127), 6–8. https://doi.org/10.1590/S0303-76572013000100001.

Llory, M., & Montmayeul, R. (2010). *L'accident et l'organisation*. Bordeaux: Préventique.

Lopes, M. G. R., Vilela, R. A. G., & Querol, M. A. P. (2018a). Anomalies and contradictions in an airport construction project: A historical analysis based on Cultural-Historical Activity Theory. *Cadernos de Saúde Pública, 34*, ePub 00130816. https://doi.org/10.1590/0102-311X00130816.

Lopes, M. G. R., Vilela, R. A. G., & Querol, M. A. P. (2018b). Agency for a systemic comprehension of work accidents and organizational anomalies. *Trabalho, Educação e Saúde, 16*(2), 1–26. https://doi.org/10.1590/1981-7746-sol00128.

Véras, R. (2014). Brasil em obras, peões em luta, sindicatos surpreendidos. *Revista Crítica de Ciências Sociais*, (103), 111–136. https://doi.org/10.4000/rccs.5559.

Virkkunen, J., & Newnham, D. (2013). *The Change Laboratory: A tool for collaborative development of work and education*. Rotterdam, Netherlands: Sense Publishers.

Chapter 10
Shared Construction of Change Scenarios for Academic Activities: The Case of a School of Public Health and Its School Health Center

Susana Vicentina Costa, Amanda Aparecida Silva-Macaia, Marco Antonio Pereira Querol, and Rodolfo Andrade de Gouveia Vilela

Abstract Brazilian higher education institutions are expected to offer teaching research and extension services to the communities they belong to. University of São Paulo (USP), one of the largest educational centers in the country, offers mainly community assistance services as extension programs. However, internal changes at USP, in addition to others involving the country's health system and higher education, led it to emphasize teaching and research activities. This context resulted in a crisis affecting academic activities of USP's School of Public Health undergraduates at its School Health Center. After successive failures to solve the crisis, Change Laboratory (CL) was chosen as a method for developing new models of collaboration. This chapter analyzes and tries to respond why CL has not passed beyond modeling scenarios to implementing changes in collaboration arrangements. Researchers revisit these institutions' context and discuss what has changed as a result of the intervention.

S. V. Costa · A. A. Silva-Macaia (✉) · R. A. G. Vilela
Department of Environmental Health, School of Public Health, University of São Paulo, São Paulo, SP, Brazil
e-mail: as.amanda@gmail.com

M. A. P. Querol
Department of Agronomic Engineering, Federal University of Sergipe, São Cristóvão, SE, Brazil

R. A. G. Vilela et al. (eds.), *Collaborative Development for the Prevention of Occupational Accidents and Diseases*, https://doi.org/10.1007/978-3-030-24420-0_10

10.1 Introduction

In the last 30 years, two major transformations in Brazilian society affected health higher education institutions: first, Brazilian Unified Health System creation (SUS[1]) (Brasil 1990) and, second, the Guidelines and Bases for Education Act (Brasil 1996).

SUS is viewed currently in Brazil as a system still in construction that suffers from chronic underfinancing, therefore having difficulties in complying with its guidelines: regionalization, decentralization, hierarchy, and social participation. More recently, SUS's underfinancing was aggravated by political and economic crises that also restrained public universities' budgets for research and health professionals training.

The crisis that hit public universities (Santos 2011) and SUS has demanded changes of models and different ways of organizing these institutions' teaching, research, and extension activities. Supported by these pillars, universities would keep close to health services and communities (Vasconcelos et al. 2016; Flores et al. 2015; Ellerry et al. 2013; Santos et al. 2000), promoting contextualized education of students, permanent education of health professionals, and improvement in integrated care of population (Ellerry et al. 2013).

Despite some examples of successful experiences of organic relationship among these activities, a real proximity between academic knowledge and practice, truly aligning research and extension axes, is still difficult. All people involved in these changes are demanded to create innovative ways of organizing university pillars (Santos 2011).

In this chapter, we deal with academic activities of research, teaching, and extension developed inside the biggest public university in Brazil: the collaboration between University of São Paulo's (USP) School of Public Health (FSP) and its School Health Center[2] (CSE). USP reacted to SUS' creation and to the Guidelines and Bases for Education Act through internal changes that emphasized teaching and research activities, leading to a crisis in the relationship between CSE, as a field for academic activities of undergraduates and graduates, and FSP. After a series of failures in solving the crisis, Change Laboratory (CL) (Engeström et al. 1996; Virkkunen and Newnham 2013) was chosen as method for developing a new model of collaboration.

At the end of CL intervention, practitioners had an expanded view about the problem, understanding that the crisis between CSE and FSP was actually a crisis between public university and society. This expansion of concepts made actors

---

[1] SUS grants free access to health services to the entire population. Its principles are integrality, universality, and equity. In its guidelines, it also prescribes the articulation of health policies with other public social policies in order to contribute for results in health and quality-of-life improvement.

[2] This School Health Center is a healthcare unit located on FSP campus where undergraduate health students assist population from its assigned area.

envision solutions to future collaboration able to offer sound academic-pedagogical relations and work processes for SUS.

> This chapter wants to answer: Why this CL did not progress from the phase of modeling solutions to their implementation?

We present results and specific learnings from CL double-stimulation technique applied during sessions and highlight the solutions' modeling phase, discussing how their activity changed after the intervention.

Section 10.2 brings the context that gave origin to intervention demand; the hypotheses built from ethnographical data collected previously, session dynamics, and the role of practitioners. In Sect. 10.3, we expose the main contradictions formulated during historical analysis sessions, in order to clarify the discussion of possibilities for activity development and the scenarios envisioned by participants as potential solutions for collaboration between the institutions. In the last section, we discuss the changes that occurred after CL and why collective learning did not last.

10.2 CL Intervention Context

10.2.1 Demands for CL Intervention

On one hand, FSP director put forward a demand related to CSE's increasing distance of its old supporting functions to teaching and research in undergraduate courses. Such activities offered a contextualized practical experience in primary healthcare but had started to break apart from USP's training and education objectives.

On the other hand, she also brought forward her concerns about deterioration signals in professional relationships, evidenced by conflicts among CSE personnel and users or reception and appointment employees, personnel complaints of an excessive workload, and increasing health leaves in these sectors. Additionally, a research group on workers' health from the FSP was looking for a place to receive a pilot project of CL. The team of researchers offered this formative intervention method as a possibility to improve work situation at the CSE and understand the Days Away from Work Cases (DAWC) problem.

After negotiating the intervention with the two institutions' directors, we used several techniques for data collection (see Table 10.1).

We invited people who were directly or indirectly involved with teaching, research, and extension activities at CSE (by the time of the intervention or before) to participate voluntarily[3] of CL sessions. Ethnographical data collection was made from 2013 to the second half of 2015. Besides, other data were collected throughout

[3] FSP Ethics in Research Committee, USP, approved this study.

Table 10.1 Summary of the methodological procedures

	Procedures	Purposes	With what? With who?	How much time?
Phenomenology phase	Face-to-face interviews	Collecting ethnographical data Elaborating assumptions about activity, main problems, historical contradictions, and innovations	Semi-structured script with CSE and FSP staff	23 interviews (approximately 17 h)
	Participant observation			13 events (approximately 22 h)
	Group interviews			
	Document analysis			
Intervention phase	Meetings	Planning CL sessions and analyzing data before and between sessions	Researcher-interventionists with audio recording and field diaries	13 meetings (18 h)
	Sessions	Achieving expansive learning actions	Double stimulation with CSE personnel and users and FSP professors and students	13 sessions (24 h)

Table adapted from Costa (2017)

the intervention, in order to understand their usual work and how social actors perceived their activity systems (AS) development.

Our initial hypothesis was that there was a single activity system (AS) integrating both institutions, FSP and CSE, and that it involved a shared object with two dimensions: "student training" and "service to users."

10.2.2 Sessions Dynamics

Data collected during ethnographical phase allowed researchers to understand the activity system and to elaborate hypotheses, serving as well as mirror data during CL intervention. There was a careful preparation process, so they could be used with the practitioners in every CL session, throughout the cycle of expansive learning (see Fig. 2.4). Each of them was intended to generate concrete outcomes and learning actions (see Table 10.2) through double-stimulation strategies (see Sect. 2.7).

From the seventh session onward, participants started modeling collaboration between activities. Due to low participation of FSP professors in these sessions, some were run separately, either with only professors or just with employees, students, and users. We adopted this strategy to try and deal with professors' historical

Table 10.2 Double stimulation during CL sessions on activities of collaboration between CSE and FSP

Session	Purpose	First stimuli	Second stimuli	Concrete outcomes	Learning actions achieved
1 and 2	Mapping and defining the problem	Actors speech extracts on a slide Question: what does it mean? Do we have a problem?	Write down a list of problems	List of problems	Recognizing a systemic problem
3	Analyzing collaboration between FSP and CSE (from the beginning)	Questions: what were the academic activities at CSE and FSP at the beginning of their relationship?	Activity system Change matrix	Change matrix for the year 1925 completed	Understanding active collaboration within the AS
4	Analyzing collaboration after 1925	Change matrix for the year 1925 completed Question: in which period should collaboration between AS be analyzed?	Timeline	Construction of a timeline	Recognizing collaboration's complexity within the AS
5	Analyzing collaboration in academic activities (teaching, research, and extension)	Timeline Question: what were the turning points in activity/ collaboration time line?	Timeline divided into management periods	Timeline restructuration	Managers influence resulted in distancing or approximation between CSE and FSP
6	Defining the turning points in collaboration	Timeline restructured	Hypothesis of activities of collaboration between CSE and FSP	Division of time line in appointed periods	Understanding the systemic process of distancing

(continued)

Table 10.2 (continued)

Session	Purpose	First stimuli	Second stimuli	Concrete outcomes	Learning actions achieved
7 and 8	Defining collaboration forms between institutions	Timeline divided into periods CSE employees speech extracts on collaboration possibilities Question: what are the concrete ways for collaboration in the future?	Matrix of scenarios	List of collaboration activities Elaboration of collaboration scenarios	Reconceptualization of collaboration between institutions
9	Defining collaboration forms between the institutions (session made only with professors)	Problem exposed on a slide List of collaboration activities Question: how should FSP's training field be organized?	Matrix of scenarios	Analysis of what a practice field has to offer in order to attend to academic needs	Identified the need for an interinstitutional PPP[a] Hypothesis of collaboration
10 and 11	Reflecting over sessions 7, 8, and 9 results and proposing collaborative activities between institutions	Scenarios and draft of the 9th session Question: what do we understand as collaborative activities?	New model/ plan for organizing the activity (that would be summarized in a letter to FSP director)	CSE practitioners decide not to lay down an unilateral document	Recognition that the model/plan and the PPP[a] should be built by both institutions with participation of FSP and CSE actors

[a]Political-pedagogical project

resistance to discuss with CSE staff about changes necessary to improve collaboration. Although we did not expect the professors' behavior, it helped us to disconfirm our initial hypothesis.

Throughout the intervention, we noticed that this was a boundary crossing CL (see Chap. 2) in which one of the main purposes is the crossing of boundaries.

Since participants' dialogue was fragile, one of the possibilities for overcoming this situation was to carry out separated sessions. The concrete outcome or learning action from one group session became a first stimulus in the other group session (Virkunnen and Newnham 2013).

10.3 Main Contradictions

In double stimulation, first and second stimuli facilitate participants and researcher-interventionists' comprehension about the systems' inner and historical contradictions (CSE's and FSP's AS). By the time of this intervention, both AS were in different development phases (Costa et al. 2018). On the one hand, FSP was in the need state phase; on the other hand, CSE was in a double-bind phase (more advanced crisis) (see Sect. 2.6).

At the health center, primary contradiction was related to a disassociation in the concept of "School Health Center": healthcare vs. teaching. When the health center was funded, according to historical analysis, both users and students were its system's objects. Some critical events during this AS development made assistance and teaching mutually exclusive:

- Creation of the SUS and municipalization of health in São Paulo state (new rules in the health system)
- USP's modification of internal rules
- Publication of a federal act establishing Guidelines for Higher Education (new rules in the teaching system)
- The contract between CSE and Municipal Health Secretariat for specialized services (focus on service)
- Changes in the age and epidemiologic profile of the population in the assigned area of the CSE (changes in the work process of the professionals)

Such events consolidated the separation between the two AS, because the institutions' activity systems ceased to share their objects and results (student/user and teaching/healthcare).

This contradiction with FSP's demands and the change of object generated secondary contradictions.

At FSP, it was possible to observe that clinical assistance stopped being academic activities' central motivation, particularly after postgraduate courses begun to be offered in this chapter in 1970 and SUS launch in 1990. Professors' strategies to teaching, research, and extension changed. Increasing pressure for research and publications led professors to work less in extension activities such as healthcare at CSE (Costa et al. 2018).

Besides primary contradictions that directly affected these two systems' relationship, secondary contradictions were also perceived, contributing to the problems observed during the intervention (Costa et al. 2018). SUS processes implementation at the health center brought up a major contradiction: a demand for integrated healthcare and the offer of specialized services. FSP professors expressed how this contradiction had an adverse effect over their mutual collaboration:

> So, for my students [...] this place has nothing exciting. [...] the way it is, it is not viable [...]. The CSE [Health Center], in fact, is not part of the professors' project anymore. It did once! The school, as a research institute, does not see CSE as a locus [of practice].

Once these central contradictions became clear, researchers understood that these institutions' relationship should overcome frontier barriers and provided complementary tools, such as asking participants about which possible scenarios could be envisaged for collaboration.

10.4 Modeling Scenarios for Possible Solutions

Seventh and ninth sessions' primary purpose was to stimulate reflection about CSE activities and to build possibilities of developing them to collaborate with FSP's needs. In two separate sessions, one with FSP professors and another with CSE's staff, users, and students, participants were oriented to draw up a draft of possibilities for this activity. In the session with CSE participants, three scenarios were offered as stimulus for discussion:

- Scenario 1 – CSE bound to FSP and São Paulo City Hall. CEAP, the support center for the School of Public Health,[4] would act as a social health organization (OSS[5]) able to recruit staff and manage financial resources.
- Scenario 2 – CSE linked to FSP and the São Paulo City Hall, with the last one responsible for selecting an OSS to manage it.
- Scenario 3 – CSE linked only to city hall, who would be in charge of selecting an OSS to manage it.

The CL practitioners discussed and rejected the third possibility straight away, arguing that this situation would change its original characteristics of School Health Center and make it impossible to preserve its link with USP. In such a scenario, they would prefer to interrupt activities altogether at the center or to close it. Table 10.3 presents first and second scenario as viewed by participants.

During professors' session, it was not possible to create collaborative scenarios, but essential elements emerged that could help CSE to become an engaging field to FSP students. In other words, CL practitioners discussed about concepts or perspectives that could reflect on a new collaboration model for these institutions. According to them, it would take:

- CSE incorporation to SUS, so that the health center could be organized as a model for integrated care, attending to specific needs of FSP courses
- The faculty to value extension activities as much as research ones, to ease production contradictions

[4] CEAP develops activities to support FSP professors in managing Public Health and Nutrition courses teaching, research, and extension programs and projects.

[5] OSS are private, nonprofit institutions hired by the state or city hall to manage SUS services.

Table 10.3 Matrix of future scenarios for CSE in its relationship with FSP

Question	Scenario 1	Scenario 2
What would change in the work process?	1. City hall finances primary healthcare 2. Staff hiring for implementing assigned area work team, dentistry team, and pharmacy reopening 3. Staff with double employment relationship (difficult to manage) 4. CEAP hires most of the staff 5. CEAP restructuration 6. Services will depend on SUS chronograms	1. Loss of work reflection space[b] 2. Increase in turnover (services loss of quality) 3. OSS mediates CSE relationship with the city hall 4. Performance and productivity assessment 5.Changes in human resources policy 6. Loss of autonomy 7. Change in working hours 8. Decrease in assigned area. 9. Health professionals do not research 10. CSE "vulnerable" to political changes 11. End of the postgraduate *lato* sensu program 12. FSP staff is minority 13. Some employees would leave CSE
What are possible alternative proposals?	1. Public Health undergraduates would practice at CSE; 2. Increased participation of Nutrition undergraduates 3. Training focused on preparing professionals to work at SUS 4. Stronger ties with FSP	1. Work Education Program (PET) 2. Since staff would change, it is difficult to imagine other programs related to undergraduate courses
What are the advantages for CSE?	1. It wouldn't be closed 2. It would have increased financial resources 3. It could adjust staff's roles to educational needs 4. It would preserve existing relationships with other education institutions	1. It wouldn't be closed 2. It would have increased financial resources
What are the advantages for users?	1. Reopening of pharmacy 2. Odontology service 3. Social control preserved 4. Increased working hours (up to 6 PM). 5. Possible new team to work at the assigned area	1. Reopening of pharmacy 2. Odontology service 3. Social control preserved 4. Increased working hours (up to 6 PM). 5. Possible new team to work at the assigned area 6. Opening of another UBS[a] at the assigned area

[a]*UBS* primary care units

[b]Space for reflection means practice of having meetings for discussion of clinical cases among the team, to define therapeutic guidelines, to guide students in the course of upgrading, or to discuss administrative and management problems

- A shared political-pedagogical project (PPP), aligned both to FSP and CSE needs, as mentioned by professors:

> [...]SUS is preponderant within the School of Public Health project. [...] SUS has some PPP which are in line with the School's political-pedagogical project but not with CSE's. Also, some SUS' PPP are in line with USP, but not with FSP (FSP professor, ninth session).
>
> This alignment of PPPs, I mean in research... You see, even in a scenario in which CSE belongs to the School of Public Health, we do not build pedagogical-political projects or shared research that reverts as potency for the service that creates innovation. We do not have it. This has to be built, and the problem is: Is it possible to do it? (FSP professor, session ninth).

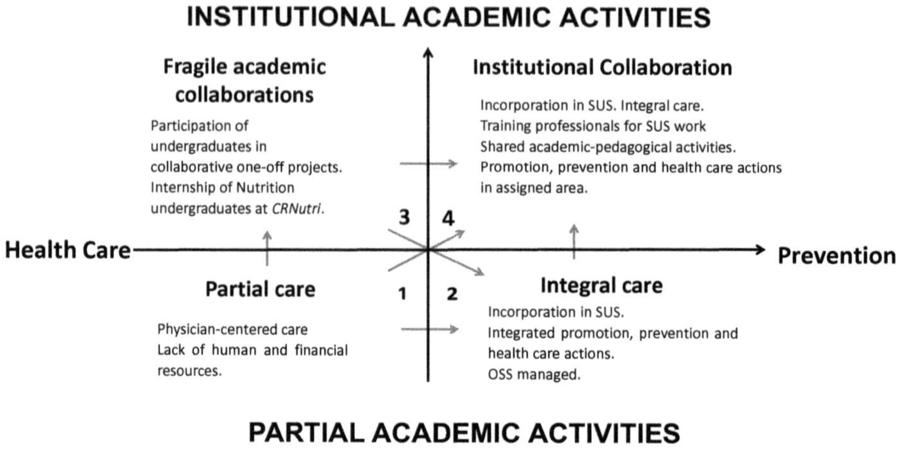

Fig. 10.1 Possibilities of development for overcoming contradictions (*CRNutri* is a Nutrition Reference Center integrated to CSE). (Adapted from Virkkunen and Newnham 2013)

Thus, participants learned that a solution for the crisis passed by the institutions rapprochement, through collaborative activities. Based on ideas from the modeling phase and clarification of central contradictions behind disturbances, researcher-interventionists were able to display a cooperative teaching-research-extension activity that could be developed into a zone of proximal development (ZPD) model (see Fig. 10.1)

ZPD modeling allowed us to present both the cooperative activity and different directions in which solutions can be organized to solve system's contradictions. In December 2016, we brought up this scheme of development in a meeting with CSE and FSP directors, as well as undergraduate and postgraduate commissions' representatives. This meeting had a double purpose: to advance toward a new AS (see Fig. 2.1) modeling phase and to get participants (in particular those from FSP) to commit to implementing a new model.

Participants discussed all possibilities of development and then decided that some actions should be implemented, in order to make collaborative activity viable. The form chosen for this collaborative activity was one desired by different CL actors (see quadrant 4 in Fig. 10.1). It was decided to establish a project group formed by CSE and FSP members to try and write down a common PPP. Undergraduate and postgraduate commission members, together with health center staff, committed to propose collaborative actions to boost FSP teaching using CSE as a practice field. A delivery time of 4 months was established for the group, after which there was to be a meeting to discuss it with both institutions directors.

However, in the meantime a more serious crisis scenario came up that went far beyond the two institutions relationship: an imminent threat of closing down CSE unsettled CL learning process deeply.

A similar funding crisis affected both USP and SUS, with impacts on their financial imbalance. At that point, CSE was USP funded while receiving small monthly installments from the city hall. FSP professors' assessment of this money transfer was that it meant an undue transfer of responsibility from the city to the university.[6] In this context and because of USP's financial crisis, university rector threatened to close down CSE; on the other hand, the municipality refused to agree on a higher budget for hiring new professionals and assuring CSE continuity. The opportunity of proposing a new model was postponed. A new funding agreement with the city hall ensured resources transfers and new employees hiring, changing it into a geriatrics specialized unit.

10.5 Discussion: Why the Cooperative Activities Did Not Concretize?

CSE employees had potential to overcome deficiencies of the collaboration model, placing education and health at the basis of a new model. However, CSE's survival was threatened by that time. At this crossroads, to keep services running for its area population was more important than academic activities, and the funds offered by city hall's contract allowed it. Survival crisis determined CSE staff's priorities. Throughout the intervention, we noticed the involving of CSE personnel in an event organization that made visible the CSE quality of work to the FSP and society.

> It seems to us that, although contradictions can push forward an AS development, it will require an object and its mediations. In this case, the object was offering healthcare to the population regardless of CSE's collaboration with FSP.

Collaboration with FSP's AS was not a problem as serious as the survival crisis. This was expressed by a participant two years after the last CL session:

> CL helped us to understand CSE better and to acknowledge other opportunities, but change did not happen! I think it was because management established new partnerships and agreements, expanding its activities and strengthening the work within its community. (FSP Professor)

It was also difficult to create a collective learning space because of FSP professors' poor participation, which prevented them from implementing what was defined during CL sessions. CL intervention was not able to engage professors into the changes, maybe because they couldn't see the point due to their perception of crisis – why to invest in a service that could die any moment?

In spite of that, both CSP and FSP staff moved on to modeling new ways of collaboration. However, CSE's staff tested a new model that solved its survival prob-

[6] USP is funded by São Paulo State taxes.

lem, but not its institutional academic actions. These advances indicate that expansive learning did happen for these actors and AS, although with diverse characteristics among them.

A clear example of learning was the rich historical analysis that rendered participants aware of the collaboration crisis origins, from CSE foundation in 1925 through different stages of interaction with FSP. We have since used this intervention as a model case to demonstrate historical analysis' potential in participatory development of work research (Silva-Macaia et al. 2016). This CL intervention also provided theoretical-methodological learning to a whole group of researcher-interventionists involved in this pilot application. It showed that CL had potential in Brazilian context.

We highlight the importance of a thorough ethnographic data collection phase for elaborating many hypotheses that contributed to a finer planning of CL sessions. It also favored bonding of the main researcher-interventionist to the participants, which helped communication and commitment. Lack of depth in this phase puts all CL potential transformation at risk. In order to keep those ties strong, each session's summary was sent to participants by e-mail. These messages fostered agency, for instance, by presenting concrete data to help historical analysis.

A particular trait distinguished this intervention: repeated and committed participation of the patients served by CSE. Postgraduate students also participated, but not as often. That actors' contribution confirmed the point of view, senses, and meanings of the social needs. It is not common in Brazil that interventionist researches involve service users in diagnosis procedures or in proposals for problem-solving.

Based on this intervention case, we conclude that CL is a tool that can promote AS development in Brazilian context. Accordingly to its materialistic-historical basis, interventions cannot dispense a structural framework analysis. In this case, problems and solutions regarding cooperative academic activities were mediated by institutional, political, and financial aspects that impaired collective construction of changes.

This study was financed by the São Paulo Research Foundation with a fellowship for postdoctoral stage of Amanda Aparecida Silva Macaia (FAPESP 2015/13301–4)

References

Brasil. (1990, 20 Sept). Lei n° 8.080, de 19 de setembro de 1990. Dispõe sobre as condições para a promoção, proteção e recuperação da saúde, a organização e o funcionamento dos serviços correspondentes, e dá outras providências. Diário Oficial da União.

Brasil. (1996, 20 Dec). Lei n° 9.394, de 20 de dezembro de 1996. Estabelece as diretrizes e bases da educação nacional. Diário Oficial da União.

Costa, S. V. (2017). *Relação entre a universidade pública e a sociedade: contribuições do Laboratório de Mudanças* [dissertation] São Paulo: Universidade de São Paulo, Portuguese. https://doi.org/10.11606/D.6.2017.tde-07072017-105154.

Costa, S. V., Macaia, A. A. S., Maeda, S. T., Querol, M. A. P., Seppänen, L. E., & Vilela, R. A. G. (2018). Change laboratory: A method for understating the crisis between public university and society. *Saúde Soc., 27*(3), 769–782. https://doi.org/10.1590/S0104-12902018170845.

Ellerry, A. E. L., Bosi, M. L. M., & Loiola, F. A. (2013). Integração ensino, pesquisa e serviços em saúde: antecedentes, estratégias e iniciativas. *Saúde Soc., 22*(1), 187–198.

Engeström, Y., Virkkunen, J., Helle, M., Pihlaja, J., & Poikela, R. (1996). The Change Laboratory as a tool for transforming work. *Lifelong Learning in Europe., 1*(2), 10–17.

Flores, L. M., Trindade, A. L., Loreto, D. R., Unfer, B., & Dall'Agnol, M. (2015). Avaliação do programa de educação pelo trabalho para saúde-PET-Saúde/Vigilância em saúde pelos seus atores. *Interface, 19*, 923–930. https://doi.org/10.1590/1807-57622014.1060.

Santos, B. S. (2011). *Universidade do século XXI: para uma reforma democrática e emancipatória da universidade* (3rd ed.). São Paulo: Cortez.

Santos, A. M., Pinho, L. M. O., Loures, M. C., & Freitas, R. A. M. M. (2000). Enfermagem em saúde pública: uma proposta articulando ensino, pesquisa, estágio e extensão. *Revista Brasileira de Enfermagem, 53*(spe), 159–164. https://doi.org/10.1590/S0034-71672000000700028.

Silva-Macaia, A. A. S., Costa, S. V., Maeda, S. T., Querol, M. A. P., Vilela, R. A. G., & Seppanën, L. (2016). O processo de análise histórica do Laboratório de Mudanças na atividade de colaboração entre uma faculdade e um centro de saúde escola. In *18° Congresso Brasileiro de Ergonomia; 2016 May 23–27; Belo Horizonte, MG, Brasil*. Belo Horizonte: ABERGO. Portuguese.

Vasconcelos, A. C. F., Stedefeldt, E., & Frutuoso, M. F. P. (2016). Uma experiência de integração ensino-serviço e a mudança de práticas profissionais: com a palavra os profissionais de saúde. *Interface, 20*(56), 147–158. https://doi.org/10.1590/1807-57622015.0395.

Virkkunen, J., & Newnham, D. S. (2013). *The Change Laboratory: A tool for collaborative development of work and education*. Rotterdam: Sense Publishers.

Chapter 11
Challenges in the Care to the Injured Worker at SUS: From the Anger Network to the Shared Construction of a Care Line

Amanda Aparecida Silva-Macaia, Mara Alice Conti Takahashi, Sayuri Tanaka Maeda, Jairon Leite Chaves Bezerra, and Rodolfo Andrade de Gouveia Vilela

Abstract In this chapter, we address the Change Laboratory (CL) conducted at the network of public health-care institutions in charge of working accident-injured workers in a town in the state of São Paulo. According to Engeström's expansive learning, this activity comprises a network of systems called "boundary crossing," as different services share aspects of the same work object. Ethnographical research revealed a fragmented health-care network, which did not have coordination between services or shared responsibilities for injured workers' cases. Based on the results of applying this Boundary Crossing CL on the development of a new health-care model, we discuss the methodological innovations and the lessons about ease and difficulties learned by the researchers during this experience. It is worthy of note to highlight the tension between the current municipal management model and the formative methodology in the context of an intervention with health professionals from many different services, with different levels of development. We conclude that such obstacles affected the participants' expectation of implementing the changes they collectively elaborated.

A. A. Silva-Macaia (✉) · M. A. C. Takahashi · J. L. C. Bezerra
R. A. G. Vilela
Department of Environmental Health, School of Public Health, University of São Paulo, São Paulo, SP, Brazil
e-mail: as.amanda@gmail.com

S. T. Maeda
Department of Public Health, School of Nursing, University of São Paulo, São Paulo, SP, Brazil

R. A. G. Vilela et al. (eds.), *Collaborative Development for the Prevention of Occupational Accidents and Diseases*, https://doi.org/10.1007/978-3-030-24420-0_11

11.1 Introduction: The Context of Municipal Health Care for Injured Workers

The municipal health system concerns a set of technologies arranged at different levels to assist its users. One of the meanings of care in this context is to welcome, treat, attend to, and respect the suffering human being – suffering which is mainly due to social fragility – with quality and resolution of his problems (Pereira and Lima 2008).

Work accident care requires, in order to achieve effective health outcomes, that specific aspects of work be addressed, such as what is done and when, how, with whom, and what type of contract. Moreover, perhaps more than in any other care, the health of workers is a field of permanent conflict between capital and labor, which determines very much the configurations of care offered and the possibilities of exercising it.

Because a single health service is insufficient to address all health problems that occur in a given population, a service network may be able to address them, provided that it is organized to provide good clinical, preventive, and health promotion results. In this sense, the integrated health services networks proposed by the Pan American Health Organization contribute to care in a comprehensive manner and directed to the specific characteristics of the citizens of a territory or region.

In this way, an integrated health services network aimed at the work accident combines aspects both common to other users of the health system, as well as specifics related to the configuration of the world of work and its biopsychosocial requirements. Such a network should be interpreted as a project in permanent construction and improvement of professional and institutional profiles, the latter derived from the performance and competence of their management.

Brazilian SUS[1] – Unified Health System[2] – includes different thematic RAS (health-care networks). Work-related health problems are dealt with at RENAST (Network of Integral Attention to Workers' Health). It is organized differently in each Brazilian state and municipality. In the municipality locus of this intervention, the network has 68 health services that should work seamlessly with CEREST – the Workers' Health Reference Center of the municipality, being ordered by the primary health care.[3]

Professionals of CEREST in the municipality, in close relation with the School of Public Health of the University of São Paulo (USP), felt the need to investigate more widely the work accidents, represented in bow tie model (Almeida et al. 2014).

[1] We chose to present all the abbreviations in the text as they are in Portuguese, translating their meaning into English.

[2] SUS is the Brazilian public health system, whose main principles are universality, equity, and integrality. It aims to ensure an interorganizational action between different services that have repercussions on the health and quality of life of individuals.

[3] Primary health care refers to the term internationally known but incorporating the principles of the Brazilian Health Reform and, therefore, emphasizing the reorientation of the care model, based on a universal and integrated health-care system.

This model is a graphical representation that brings to the center the critical event – the work accident, on the left side the proximal and distal causes that culminated in its occurrence and on the right side the consequences, equally proximal and distal, such as lesions and physical, emotional, and social sequelae to injured workers. Damage mitigation systems are differential elements in the scenario that makes up the right side of this analysis matrix (Dianous and Fiévez 2006). In this broader view of work-related accidents, health interventions for the injured have the potential to avoid further aggravation and reduce the damage of work and life incapacity for the injured (Hale et al. 2007).

The empirical hypothesis for the investigation was that this municipal RAS did not offer an integral care to the injured, since it did not intervene in reduction of disability damages. It was assessed that, although the network offered a significant range of services and guaranteed universal access to formal and informal workers, there were important assistance gaps that resulted in worsening health conditions and comorbidities, with high costs for the public health system.

Based on this need that had its origin within the service and was further built in dialogue with the university, the demand was formally established in 2011 as a subtheme on the thematic project "*Work accident: from socio-technical analysis to social construction of changes*," supported by FAPESP (São Paulo Research Foundation). In this chapter, the innovations and the methodological limits seized by the researchers in this experiment are discussed, in order to present the obstacles to the implementation of the innovations in the activity and to its sustainability possibilities in face of the management model adopted in the municipality.

11.1.1 The Construction of the Demand for the Intervention: Outlining the Learning Process

To initiate the process and build a demand, a discussion group was created gathering CEREST, MTE (Ministry of Labor and Employment), MPT (Labor Prosecution Office), CMS (Municipal Health Council), and workers unions representatives, where the research proposal was recognized as an important public health issue of social relevance to the municipality. A pre-diagnosis was carried out as a postdoctoral project at the USP, which pinpointed a network fragmentation; a transitory presence of the accident victims in all the units where they passed; lack of follow-up of the injured through the services since no professional was assigned to that; and gaps regarding social security protection and physical, psychological, social, and professional rehabilitation of these workers.

These relevant conclusions about the care of the work accident generated needs for intervention in the network. However, even though several professionals were interviewed, diagnosis was made in a traditional way, that is, from the perspective of the researcher (Takahashi and Cerveny 2016). Thus, in order to provide sustainable changes that could account for integrality in health as the full application of the reception, bonding, accountability, and resolution devices (Paim 2012), it was necessary to construct a new participatory diagnosis engaging the point of view and knowledge of service and management professionals.

It was considered that an intervention in the care network to the injured worker – since it is formed by a set of different services where each member forms a singular system of activity which shares dimensions of the same object, e.g., the same users – demanded the application of the Boundary Crossing CL model (Virkkunen and Newnham 2013; Seppänen et al. 2016).

In this chapter, we analyze data collected in Boundary Crossing CL sessions and in the researchers' field diary on the meetings with municipal health management representatives. Part 2 presents the general characteristics of this intervention: the distribution of the attendance of the various participants of different activity systems throughout the phases of the cycle and a historical line of the intervention itself, in order to give visibility to the challenges of a Boundary Crossing CL.

In Part 3, the results of the process are described, considering that the cycle of expansive learning can be divided into three phases – questioning, analysis, and modeling – with the predominance of different types of learning activities in each of them and without linear chaining (Virkkunen and Newnham 2013). It highlights a double-stimulation sequence during the modeling of the care line which resulted in a model created by the participants to establish new care for the injured. Part 4 discusses the limits and sustainability of the methodological innovations, based on the data from the CL sessions and from the researchers' field diaries. This subsection addresses the municipal management's role in the choice of participants for the testing of proposed solutions for implementing a new model of care for the injured.

11.2 The Boundary Crossing Change Laboratory in the RAS

The specificity of this type of CL lies in its unit of analysis, which focuses on the collaboration of different activities of different services that serve the same users, who are, for each of the systems involved, part of a broader object (see Chap. 2). It occurs the crossing of the boundaries of the services, so that the objects of the activities are shared, and these activities are systematically carried out together (Virkkunen and Newnham 2013).

In the case of health care, boundaries are created mainly by medical specializations and bureaucratic restrictions that can fragment services and their outcomes, making the process of care difficult (Kerosuo 2008). In fact, the fragmentation of services is a historically constituted reality that generates segmented care and decreases the power of the structural dimension of combining resources, as well as guaranteeing the satisfaction of users' needs. This context does not favor the systemic production, which should be desired as a RAS' characteristic (Mendes 2011).

In the studied municipality, fragmentation occurs also within the services, through a twofold management: each service is administered concomitantly by medical coordination and nursing coordination. In practice, the double coordination

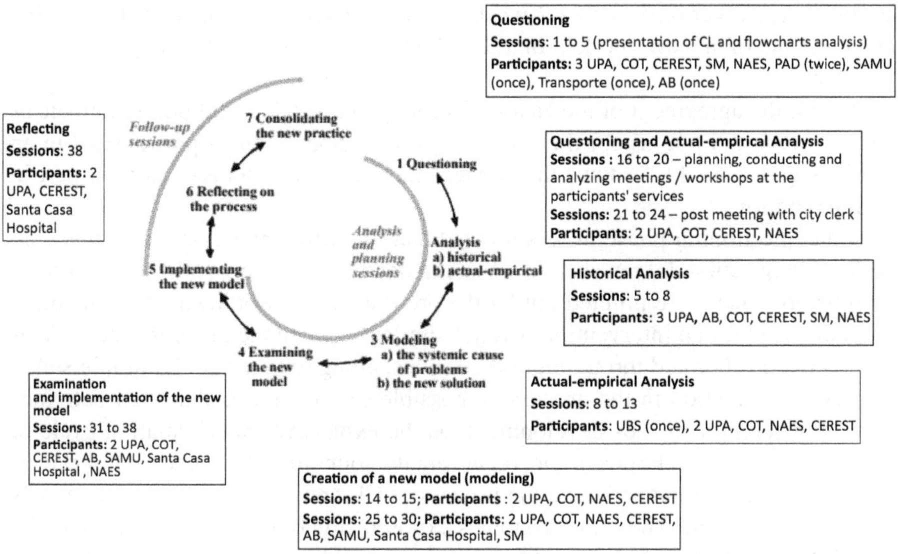

Fig. 11.1 Dynamics of the participation of the activity systems in the Boundary Crossing Change Laboratory phases. (Adapted from Virkkunen and Newnham 2013). Caption: UPA – Emergency Care Unit; COT – Orthopedics and Traumatology Center; CEREST – Workers' Health Reference Center; SM – Psychosocial Support Center (mental health service); NAES – Health Education Center; PAD – Home Care Program; SAMU – Mobile Emergency Service, AB – Primary Health Care

reflects the theoretical and practical difficulties to break traditional dichotomies in the coordination of command activities and creates a difficulty for more effective interdisciplinary management. Fragmentation of care was also perceptible in the relationship between services, including those from different complexity levels (primary, secondary, and tertiary). This could be seen by the bureaucratic communication (using to and from forms and other local forms sent via users) and the lack of direct communication between services' team members or cases' discussion meetings and joint planning of integrated therapeutic projects.

In view of the various services involved in the injured worker care and the way they were organized, the proposed intervention in the RAS, prepared by the university and CEREST, was presented to the Municipal Health Secretary and to the Municipal Health Council. Once the implementation of the CL was agreed upon, the next step was to invite services' medical and nursing coordinators integrating SUS' three-level care network in town[4] who took part in the therapeutic itinerary of the injured workers, according to the pre-diagnosis elaborated. Soon after, the

[4] Emergency Mobile Service, Hospital Vacancies' Regulation Center, Basic Health Unit and Family Health Program, Emergency Care Unit, Orthopedics and Traumatology Center, Specialty Center, Municipal Physiotherapy Center, Psychosocial Care Center, and Workers' Health Reference Center. It was planned to invite to be integrated in the second session: fire department, hospitals, center of specialties, and eye clinic.

sessions were organized, with a 2-week interval between the stages of questioning and creation of a new model. In the following phases, the meetings occurred monthly.

Despite the agreement of the Health Secretary, the services did not participate as desired: some did not accept the invitation to participate, others participated in an erratic way. Figure 11.1 shows the services (activity subsystems) that took part in the intervention.

In the questioning phase, it was noticed that the different services that made up the RAS in question did not share the same understanding of causality and solutions for problems that were pointed out by the pre-diagnosis. Moreover, the perception about the need for an intervention was not similar between the different actors. This was being corroborated throughout the sessions, along with the oscillation in stakeholders' participation in the process. A possible explanation is that their activities were in different stages of development, as the expansive transformation cycle of activity (see Fig. 2.4). For example, regarding the four emergency care unit services, three nursing coordinators began the process of CL and only two were assiduously present until the end. These professionals exposed situations of daily work that disturbed them and provoked them to analyze the problems in and between the network's activity systems. The actions they developed for the benefit of the users characterized personal initiatives to remedy certain deadlocks; however, they were legitimate professional actions inherent to health care.

On the other hand, professionals from the physiotherapy center or the transport sector, for example, did not express dilemmas or conflicts regarding the attention to the injured worker or the relationship with the other services of the RAS and, therefore, were not willing to participate of intervention, whose principle is the voluntarism of its members. The coordinating physicians were also absent along the CL. As one of the characteristics of the services in the municipality was the division of administration between medical coordination and nursing coordination, this absence had negative significance for the process, because it did not allow an alliance between manager and participants that would carry the participants' belief that the proposed changes would actually be implemented.

There were 38 sessions between April 2016 and July 2018 and several planning, analysis, and negotiation meetings. Figure 11.2 highlights the events that, for the researchers, were relevant throughout the intervention as driving elements in the process of expanding the object of the work activity and that determined new directions in the LM.

The first moment chosen was the recognition of the social necessity of the "provisional learning activity," that is, of the intervention with the Boundary Crossing CL. At this point the process was started, with an apparently shared object between the intervention system and the system management system, which granted permission for the development of the intervention.

A better appropriation of the methodological tools by CL participants is represented by the second relevant moment. This methodological innovation will be reported in more detail in the fourth section and refers to workshops mediated by the CL participants themselves in some of the network's services. The objective was

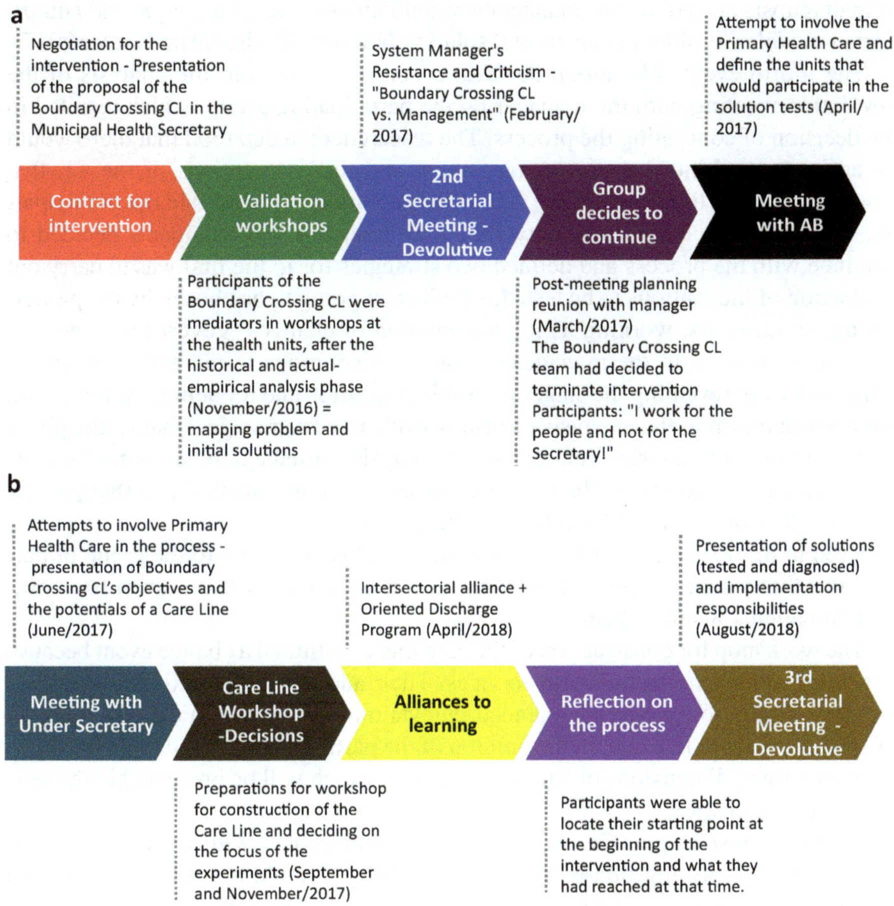

Fig. 11.2 Events that caused Boundary Crossing Change Laboratory changes, according to researchers. Caption: CL - Change Laboratory; AB - Primary Health Care

to validate the diagnoses built by the group during the questioning and analysis phase, and also to observe the views of other health professionals about changes in care for the injured in the municipality. The materials provided by this activity were the basis for preparing the first return dialog to the central management system.

The third moment was the devolution meeting, which was a stimulus to one of the main changes in the CL's course. The meeting exposed a crisis between the provisional learning activity and the municipality's management activity. It was evaluated that at that moment there was no sharing of objects between the two activities, which had different concepts and strategies for the construction of innovative solutions in relation to the observed disturbances. While the CL has its theoretical and methodological basis in a participatory approach in the construction of both diagnosis and solutions, management guidelines revealed a unique and authoritarian model of problem analysis and solution building (Bodrožić and Adler 2017).

The fourth event, *The group decides to continue*, refers to the analysis of the devolution meeting with the manager by the participating group, which resulted in the decision of continuing the process. The researchers understood that there would be an incompatibility between management and the assumptions of the CL that would undermine the construction and implementation of changes in the municipality. Therefore, the learning activity should be interrupted. Participants decided to continue with the process and defined two strategies for it: the first was to carry out a selection of innovations to be tested, which were going to be chosen by the participating coordinators, working as implementation of changes' governance; the second was to present to the management only the solutions tested and in progress, refraining from revealing the stages of problematization and modeling. At this point, there was a return in the expansive learning cycle to earlier stages, that is, the phase of questioning, of historical and empirical analysis started again, allowing for new, more mature reflections on the previous stages and proposals built, so that participants could move forward from the new decisions.

Even in the absence of PHC representatives, it became clear to the group of participants that it was not possible to plan for a new care activities system for work-injured patients without them.

The workshop for construction of the care line constituted a change event because new integrants came from health services. Their addition to CL existing members enriched discussions about the planned innovations and helped in the development of something more feasible, in the opinion of the participants. They decided to focus on testing three dimensions of a new care model, which will be presented in the next section of the chapter.

When the test begun, CL went on (monitoring phase) with monthly sessions through the monitoring phase. We were searching to build a space for continued learning, to support participants after researchers left. During this period an intersectoral alliance was established that had "qualified discharge" as the main interface. This meant an essential part of the care line for the patient (see below). The alliance meant a more institutional space in town than CL.

To end the use of CL sessions as a provisional learning activity, the group reflected about its process. To promote participants' reflection, a two-dimensional model was offered as secondary stimulus. It was a figure in which two axes determined four quadrants that represented the concepts that had to be mastered, according to the analyses of the group. Participants were able to locate their starting point at the beginning of the intervention and what they had reached at that time.

In the third meeting, the health secretary understood that it was a final presentation that brought in the end of CL. In it, researchers presented for discussion the tested aspects of the dimensions chosen to the new model and suggested possible actions for continuity, attribution of responsibilities, and, finally, suggestions of changes that hadn't been tested because they required collaborative negotiations, for instance, to grant effective participation of PHC representatives.

11.3 Building Innovation Proposals for Care of the Injured Worker: Main Results for the Expansive Learning

According to the assumptions of the CL, all phases are constructive and require a collective maturing of the participants from the perspective of work. The proposal-building phase is highlighted, where the participants understood the methodological path proposed by the Boundary Crossing CL and developed a more systemic, object-oriented view.

The stage of problems analysis was followed by the third phase of the learning cycle, when three solutions were identified which could minimize the effects of disarticulated care for the injured:

- To structure a line of care for the injured worker articulated to the Oriented Discharge Program
- Qualitative improvement of SIVAT (Occupational Accident Surveillance System)/CEREST
- Implementation of a permanent education nucleus

A public system characterized by not-coordinated health services predominates in Brazil. As a result, even if their staff has an abstract idea of an orderly functioning of the system and how to gain access to them, users face barriers, either of misunderstanding or of unavailability of care at the services.

Another concept that was relevant to the absorption of professionals is the concept of "work accident." The research group assumes that an occupational accident is an avoidable event, but, when it occurs, it is critical in the lives of workers, an acute health need in emergency services. Their unfolding into suffering and sequelae will be treated in primary care units; however, they cannot be detached from the main occurrence.

This is in line with the results of previous researches, which showed that the demand for care of the accident victims first came into health services mainly through emergency rooms, according to the severity of health problems and the rescue services.

11.3.1 Structuring of a Care Services Line for the Injured Worker Articulated to the Oriented Discharge Program

The first proposal for care line building (Mendes 2011) is to structure it by themes, according to health problems and needs. In case of work accidents, it is important to classify them by severity in mild, moderate, and severe, since there are around 10,000 work accidents/year in the municipality studied – in an economically active population of 192,936 workers, according to the latest demographic census (Marinho 2016; IBGE 2010). For the exercise of discussing and developing a proposal for a care line, seriously injured workers were taken as model, since there was empirical knowledge that these patients require a very complex care supported by all the technologies available in the health and social security system – a clear justification of network care.

In the process of services line modeling, professionals suggested that it should be articulated with another instrument – Oriented Discharge Program – which had been developed by two hospitals in the municipality along with some primary care professionals in an unstable partnership. It aims to ensure continuity of health care, as well as to foster communication with the family to prepare discharge and post-discharge. An instrument called "discharge plan" was implemented: a detailed care report for hospital discharge, aiming at generating communication between technical team and the user and his/her family and, at the same time, making care quality and safety evaluation possible.

By articulating the Oriented Discharge Program to the CL, it was enriched conceptually throughout actions of questioning, testing, and implementation, changing the course of its development.

11.3.2 Qualitative Improvement of SIVAT/CEREST

SIVAT, which has been operating continuously and systematically since 2003 in the municipality's CEREST, centralizes epidemiological information on work accidents, prioritizing intervention in sentinel events, such as serious injury and fatal accidents involving minors under 18 years (Vilela et al. 2018). SIVAT incorporated knowledge from scientific evidences, working in close partnership with researchers. It was proposed that the first input of data during care included qualitative information, homogeneous severity criteria, that would be complemented afterward by data on the continuity of care. This change will allow monitoring of care services line.

Other recommendations for improving SIVAT were (1) enabling all care units to issue a RAAT (Injured Worker Attendance Report), increasing the coverage, and storing contact data; (2) allowing multidisciplinary team members to issue this report, thus overcoming the physician-centered model; and (3) a plan of *continued education* so that emergency services professionals realize the importance of the information they provide in preventing new occurrences.

11.3.3 Implementation of a Permanent Education Nucleus

Continued education was indicated as a sustainability mechanism for the health system, in the perspective of producing social values to reinforce health surveillance and care humanization and of becoming a space that facilitates and promotes network articulation.

There was a consensus in the group that a conceptual and structural transformation of the former health education nucleus was necessary: it was formed by only one nurse, whose activity was centered on receiving and solving individual complaints, with no guarantee of equity or institutional learning resulting in changes in the care model. Continued education should derive from daily services. Also, there is a misconception, on the part of management, that educational activities are not part of the job.

The idealized solution – the implementation of a permanent education nucleus – foresees a partnership with other institutional actors, for example, a negotiation with the School of Public Health of USP to structure a nucleus of Permanent Education in the city.

11.4 Methodological Innovations: Facilities and Limits

Boundary Crossing Change Laboratory modality proved to be a useful tool to improve services provided to patients who suffer from serious consequences of work accidents. Helsinki School has produced wide literature pointing its learning potential and scope when carried out at the intersection of cross-border professional organizations or cultures... Even so, its use in a Brazilian context represented a creative effort requiring new conceptual resources and practical tools.

Integral care, as advocated by SUS' principles, cannot be offered by isolated services, even if in isolation they perform well (Pinheiro et al. 2007). Such care requires an articulated ordering between different services, able to attend to each case's singular needs – different technologies, primary care, specialized referrals, and hospital care, as well as rehabilitation practices.

The CL method itself provides an innovative and advanced theoretical framework for so-called qualification practices in Worker's Health, traditionally approached in traditional one-way specialist learning. The concept of expansive learning that guides the CL conduces to learning "something that is not yet there" for any participant (Engeström and Sannino 2010), even researchers who have prior knowledge about the subject. It is new knowledge, amalgamated by interaction and collaboration, built in a shared way.

We chose to use flowchart analyzers (Merhy et al. 1997) for mirror data during double stimulation, both in the questioning phase and as a tool during new model's testing. This artifact, which represented real cases' pathways within local health system, was useful in helping participants to discuss different logics inbuilt in care services model. Flowcharts made problems, tensions, noises, and contradictions among health network's different services easy to recognize. Services users' path through services portrays the objects produced by each of them and at the same time how different and contradictory visions permeate these same objects. In general, each service had mastery upon its activity's object and was unaware of other services' object, leading to tense relationships in what one of the participants called the "anger network."

Another powerful resource – a dramatization of a severely injured worker first assistance – rescued by SAMU and taken to UPA's emergency service was used as a double stimulation to help in identifying the systemic character of internal and external contradictions that led to the "anger network." It also helped to see systemic solutions that could lead to overcoming these antagonistic and constraining forces that operate in day-to-day services and in their many borders. Solutions' building is complex and involves new tools and approaches created in a collaborative way, in new patterns of activities, which require negotiation and governance for change implementation.

The reasons noticed for a low participation in CL sessions – the number of participants was always below expectations and did not remain stable – are related to the management model of municipal health services, financing units per procedures completed, and keeping reduced teams of professionals. In this scenario, work absences are always very difficult. These conditions led the researchers to adopt a strategy of transforming the total group of participants into a joint planning group with interventionists. Therefore, planning also became a learning action, through the realization of the workshops in different health services, supported by the researchers, as well as in the preparation of the contents for the workshops.

On the other hand, it was very difficult to disseminate the ideas for change to the services frontline team, since they could not leave their functions to take part in any activities due to high demand of users, personnel deficit, and payment per procedures:

> It is an illusion to think that we can spread information [of what happens here] ... the coordination is always very busy ... there is a patient who needs to be intubated ... the waiting room full of users [patients] waiting … an ambulance coming in ... and people cannot spread all this information, it is very difficult for someone to pass on to the others... in fact, information is mainly concentrated on some people (a nurse, coordinator and participant in the sessions)

The best dissemination of information occurred with the strategy of decentralizing the sessions to the services, which allowed the validation of systemic problems perceived, the initial solutions thought, and the emergence of new solutions.

In the implementation phase of the new model, a long workshop (twice as long as ordinary CL sessions) was held. We thought it was strategic to do so, in order to review CL process. A panel exhibition was organized, showing its path, elements from SUS network time line, identification of its main problems and their origins (first stimulus), as well as the idealized conceptual solutions (second stimulus). After a circular dance, three categories of solutions (or dimensions of the new model) and their respective instruments of change were outlined. The categories of solutions reported in the previous section were chosen based on agency and governability self-evaluated by the group. This exercise helped participants and researchers to move cognitively from abstract to concrete and to experience the new model's configuration.

However, despite the positive impact of the stimuli, this reduction to a prototype of three-dimensional solutions, rather than more comprehensive and consistent proposals built throughout the learning process, was probably due to the centralizing profile of municipal health management. There was significant expansive conceptual development during the shared construction of the care services line; however, it was abridged by low change capacity at the services level.

The process of implementing the new model was led by CEREST, the only service that had autonomy and manifested interest. Its agency results from a more stable situation, since this service experienced an operations remodeling process through CL methodology (see Chap. 2) that implied in intersectoral action within health services network (Vilela et al. 2018).

Finally, looking at work practices through the theory on modes of collaboration (Engeström, 2008, as cited in Virkkunen and Newnham 2013), we perceive that in CL's initial moment, *coordination* was the dominant mode of collaboration: each service's players acted on its object separately, according to roles and sequences of actions defined by a tacit script.

During CL, at times and for some questions, a mode of collaboration closer to *cooperation* came up: the actors involved focused on a shared object and sought joint solutions to identified problems that went beyond the established road map.

We expect that, as the changes implemented by adopted solutions become consolidated, cooperation stabilizes and a *communication* collaboration mode is reached, where actors willingly transform their roles and interaction becomes constant around a truly shared object. If this is to happen and this type of collaboration be carried out, it is urgent and necessary that PHC is made part of injured workers care services line.

11.5 Final Considerations

One of CL methodology's central nuclei is the search for reality transformation through organizing human capacity instituting forces in work processes. It is an innovative way of combining researchers and organization agents' action in mutual collaboration in order to apply investigative and intervention methodology for problem-solving and transforming the studied object. Helsinki School's proposition, widely referenced, is understood as a management innovation process by this research group.

In this retrospective analysis, besides having these characteristics, CL helped to focus on health network interinstitutional relations among many different services oriented to producing integral care for work-injured patients.

Despite the obstacles mentioned, such as the unstable participation of the players, difficulties in disseminating information to frontline professionals, and reduced governability due to the Taylorist profile of management control, we consider that expansive learning was leveraged by sustained intervention, which preserved the endogenous process of continuous creation mediated by the methodology's fundamentals. Participating actors proposed and achieved test actions of the new model. They planned the continuity of implementation with the support of other professionals.

In the conceptual field, health-care meaning and integrality principle were discussed in the sessions as a social phenomenon and network care as the one that can achieve social answers. In the historical-cultural approach, the formation of concepts is essential for the process of creation, development of knowledge, and instigating learning (Serpa and Falcón 2015).

Expansive learning, in particular for the group of researchers, who headed the formative intervention, was developed intensively during the planning and evaluation stages of the sessions throughout the cycle. The formulation of methodological steps allowing strategies' adjustment, the use of expansive learning theoretical corpus, organization of the collected information, as well as sessions' evaluation were assimilated in an interaction and real experience climate, fundamental to enable participants to recognize contradictions and formulate systemic solutions.

There is a challenge for our research group and to the service professionals who have taken part in CL: supporting their results and disseminating them throughout the network, in order to win the adhesion of all health professionals for its implementation. For the near future, it is necessary to overcome the barriers of the direct implications of the current form of municipal management, since the rigidity of the hierarchical work diminishes the autonomy and the creativity of the professionals, a situation incompatible with the development of the leveraged expansive learning.

It is urgent to overcome the Taylorist model of control over resources for a structural transformation able to offer greater autonomy to care professionals and to create end activity systems that articulate transformative care production as recommended by Brazilian Health Reform.

It seems to us that professionals' continued education is vital for services if attention models are to be organized to give visibility to care. It will require training actions beyond technology-centered knowledge, aiming at developing skills that help professionals to absorb new paradigms on organizational learning.

This study was financed by the São Paulo Research Foundation with a fellowship for postdoctoral stage of Amanda Aparecida Silva Macaia (FAPESP 2015/13301-4).

References

Almeida, I. M., Vilela, R. A. G., Silva, A. J. N., & Beltrán, S. L. (2014). Model of analysis and prevention of accidents - MAPA: Tool for operation health surveillance. *Ciência & Saúde Coletiva, 19*(12), 4679–4688. https://doi.org/10.1590/1413-812320141912.12982014.

Bodrožić, Z., & Adler, P. S. (2017). The evolution of management models: A neo-Schumpeterian theory. *Administrative Science Quarterly, 63*(1), 85–129. https://doi.org/10.1177/0001839217704811.

Dianous, D. V., & Fiévez, C. (2006). ARAMIS project: A more explicit demonstration of risk control through the use of bow-tie diagrams and the evaluation of safety barrier performance. *Journal of Hazardous Materials, 130*(3), 220–233. https://doi.org/10.1016/j.jhazmat.2005.07.010.

Engeström, Y. (2008). From teams to knots: Activity-theoretical studies of collaboration and learning at work. Cambridge: Cambridge University Press.

Engeström, Y., & Sannino, A. (2010). Studies of expansive learning: Foundations, findings and future challenges. *Educational Research Review, 5*(1), 1–24. https://doi.org/10.1016/j.edurev.2009.12.002.

Hale, A. R., Ale, B. J. M., Goossens, L. H. J., Heijer, T., Bellamy, L. J., Mud, M. L., et al. (2007). Modeling accidents for prioritizing prevention. *Reliability Engineering and System Safety, 92*(12), 1701–1715. https://doi.org/10.1016/j.ress.2006.09.025.

IBGE- Instituto Brasileiro de Geografia e Estatística. Censo demográfico de 2010. https://censo2010.ibge.gov.br/resultados.html Acessed 3 April 2019.

Kerosuo, H. (2008). Putting the patient in the middle: Managing chronic illness across organizational boundaries. In R. Sorensen & R. Ledema (Eds.), *Managing clinical processes in health services* (pp. 73–86). Chatswood: Elsevier.

Marinho, C. S. (2016). *Consequências econômicas com acidente de trabalho: realidade do município de Piracicaba* [dissertation]. São Paulo: Universidade de São Paulo; Portuguese. https://doi.org/10.11606/D.&.2016.tde-31082016-154645.

Mendes, E. V. (2011). *As redes de atenção à saúde* (2ª ed.). Brasília: OPS.

Merhy, E. E., Chakkour, M., Stéfano, M. E., Santos, C. M., Rodrigues, R. A., & Oliveira, P. C. P. (1997). Em busca de ferramentas analisadoras das tecnologias em saúde: a informação e o dia a dia de um serviço, interrogando e gerindo trabalho em saúde. In E. E. Merhy & R. Onocko (Eds.), *Agir em saúde: um desafio para o público*. São Paulo: Hucitec.

Paim, J. (2012). Modelos de atenção à saúde no Brasil. In L. Giovanella, S. Escorel, L. V. Costa, J. C. Noronha, & A. I. Carvalho (Eds.), *Políticas e sistema de saúde no Brasil* (2ª ed., pp. 459–491). Rio de Janeiro: FIOCRUZ.

Pereira, I. B., & Lima, J. C. F. (2008). *Dicionário da educação profissional em saúde*. Rio de Janeiro: EPSJV.

Pinheiro, R., Ferla, A., & Silva, A. G., Jr. (2007). Integrality in the population's health care programs. *Ciência & Saúde Coletiva, 12*(2), 343–349. https://doi.org/10.1590/S1413-8123 2007000200010.

Seppänen, L., Takahashi, M. A. C., Cerveny, G. C. O., & Vilela, R. A. G. (2016). Laboratório Interfronteiras: uma ferramenta possível para a integralidade no atendimento aos acidentados de trabalho. In *18° Congresso Brasileiro de Ergonomia; 2016 May 23-27; Belo Horizonte, MG, Brasil*. Belo Horizonte: ABERGO. Portuguese.

Serpa, G. R., & Falcón, A. L. (2015). La formación de conceptos: una comparación entre los enfoques cognitivistas y histórico-cultural. *Educação e Pesquisa, 41*(3), 615–618. https://doi.org/10.1590/S1517-9702201507135042.

Takahashi, M. A. C., & Cerveny, G. C. O. (2016). Impacto social dos acidentes de trabalho no município de Piracicaba (SP): intervenção e busca compartilhada pela integralidade em saúde. In F. Asensi, R. Pinheiro, & P. A. Mutiz (Eds.), *Bioética, trabalho e educação em saúde* (pp. 67–77). Rio de Janeiro: Multifoco.

Vilela, R. A. G., Jackson Filho, J. M., Querol, M. A. P., Gemma, S. F. B., Takahashi, M. A. C., Gomes, M. H. P., et al. (2018). A expansão do objeto da vigilância em acidente do trabalho: história e desafios de um centro de referência em busca da prevenção. *Ciência & Saúde Coletiva, 23*(9), 3055–3066. https://doi.org/10.1590/1413-81232018239.21952016.

Virkkunen, J., & Newnham, D. S. (2013). *The Change Laboratory: A tool for collaborative development of work and education*. Rotterdam: Sense Publishers.

Chapter 12
Change Laboratory in an Urban Cleaning Company: A Dialogue with Female Street Sweepers

Marina Zambon Orpinelli Coluci, Bianca Gafanhão Bobadilha, Ana Yara Paulino, and William da Silva Alves

Abstract The aim of the present chapter is to describe the experience of urban cleaning workers, mainly women, who work sweeping streets, and how the Change Laboratory (CL) method influenced the development of improvements for the workers themselves and society at large. The work performed by sweepers is extremely relevant for society and the public administration—public health in particular—because they collect the waste produced daily to maintain cities clean and minimize the occurrence of diseases. Several studies described the vulnerabilities to which these workers are exposed along the working hours, their invisibility, and social prejudice. The CL method is based on theories for collective development of ideas. In the present case, a representative group of street sweepers participated in sessions with mediators to discuss and analyze the street sweeping activity from its beginnings to the present day. Based on ethnographic data and considering difficulties exhibited by the group, the mediators had resource to double stimulation and illustrative materials to facilitate discussions. The group designed an activity system for the sweeping activity, which helped them understand all the aspects involved. Seeking better socioenvironmental sustainability, improvements in the working conditions, and prevention of occupational diseases and work accidents, the participants suggested transformative and collective solutions to be implemented in the near future.

M. Z. O. Coluci (✉) · B. G. Bobadilha · A. Y. Paulino · W. S. Alves
Department of Environmental Health, School of Public Health, University of São Paulo, São Paulo, SP, Brazil
e-mail: mazorpinelli@yahoo.com.br

R. A. G. Vilela et al. (eds.), *Collaborative Development for the Prevention of Occupational Accidents and Diseases*, https://doi.org/10.1007/978-3-030-24420-0_12

12.1 Introduction and Background

Urban cleaning is essential for society, since its absence would result in several public health problems. Nevertheless, this activity sometimes does not receive the social recognition it deserves. In addition to handling garbage, which task is a source of social exclusion, urban cleaning workers are exposed to instability and organizational and employment changes, which contribute to cause apprehension and discomfort (Gomes and Oliveira 2013).

Some of the reasons to seek this kind of job are difficulty to find other positions and fear of unemployment (Lopes et al. 2012). Globalization and production restructuring helped increase the precariousness of work, with consequent spread of insecurity and uncertainty within and outside the world of work (Seligmann-Silva 2011). These aspects intensify the fear of unemployment and the difficulties to take a stance vis-à-vis a top-down and authoritative organization or work.

Several studies pointed out the vulnerabilities to which urban cleaning workers are exposed (Costa 2002; Lopes et al. 2012). In addition to social prejudice—these workers are commonly known as "waste personnel" and "waste collectors" (and sometimes are literally treated as garbage)—they are also exposed to poor working conditions likely to cause work-related problems, such as musculoskeletal pain, cuts, and poisoning (Lopes et al. 2012). Urban cleaning workers perform their work even when feeling pain, and the rate of low back pain was found to be higher among workers exposed to psychosocial demands compared to the ones who are not (Pataro and Fernandes 2014).

The sociodemographic profile of this population is quite particular: most urban cleaning workers have a low educational level and earn about the minimum wage (Gemma et al. 2017; Bandeira and Almeida 2015). Their poor working conditions do not only affect their physical health but also their psychological health and social recognition (Costa 2002; SIEMACO 2011).

Formative interventions with this population of workers are necessary to at least minimize the suffering derived from the contradictions inherent to the organization of work and help them achieve social respect.

The decision to apply the Change Laboratory (CL) (Engeström et al. 1996; Virkkunen and Newnham 2013) methodology in the present study was based on the assumption it could contribute to the formulation of a new activity system for the urban cleaning service. Aiming at achieving socioenvironmental sustainability, the new system ought to include improvements in the working conditions, prevention of occupational diseases, and work accidents.

The Workers' Health Reference Center (CEREST) of a city in the state of São Paulo, Brazil, with an estimated population of 397,000 inhabitants in 2017, analyzed the growing rates of sickness and accidents among public cleaning workers and implemented specific measures targeting their health and safety at work. The initiative, however, had poor success, and being aware of the need for a systemic study, CEREST called the CL research group at School of Public Health, University of São Paulo (FSP/USP), to conduct a systemic formative intervention—the trigger was the hiring of a new urban cleaning company for a period of at least 20 years.

12.2 Negotiation

A meeting following the CL session model was held by the company managers, CEREST representatives, and CL investigators. Double stimulation was used to promote group discussions. Pictures of the activities currently performed by workers were shown (first stimulus) followed by questions on how this type of work could be performed better (second stimulus). From that point onward, the discussion focused on the establishment of an object to be analyzed by both workers and managers leading to new perspectives for improvement of the work process.

Following analysis of the proposal by the company, the initial demand shifted from application of CL to waste collectors to street sweepers.

12.2.1 The Sweeping Activity

The sweepers at the analyzed company are mostly women. This fact brought other social demands into light.

The situation of women as general service providers has some particular aspects: "their condition as workers without formal qualifications and submission to the discipline and logic of capitalist relations, essentially as concerns the availability of the use of this workforce by capital, such as their female condition, whose roles are characterized by subordination to male dominance relationships manifested in the home environment or wage labor" (Chillida and Cocco 2004). In addition, women face a double burden, and despite the many women's struggles, they are often condemned to unemployment (Monteiro-Cocco 1997).

Sweeping is always performed in pairs; each pair is responsible for sweeping the curb of streets and collecting daily waste, covering about 2 kilometers every day, on average. Their working tools are a cart, brooms, shovels, trash bags, and personal protective equipment (visibility clothing, shoes, and gloves).

The working day starts at 6:00 at a place denominated "concentration," where the working tools are stored. Then the sweepers go to the allocated worksites established according to a map drawn by the employer. They are expected to return to headquarters at about noon for lunch and rest. The working day finishes at 14:00.

It should be noticed that as the company had been hired for a very long period of time, some changes were made during the period of data collection. Changes affected the number of sweepers, supervisors, and working hours. Due to municipal budget issues, the amount of money paid to the cleaning company decreased, resulting in reduction of the number of sweepers and supervisors. The working day, which previously finished at 14:00 and included a meal break from 11:00 to 12:00, shifted to six uninterrupted hours starting at 6:00, with a meal break from 12:00 to 13:00, followed by a 1-hour rest break at headquarters, the working day finishing at 14:00. Under the previous schedule, the workers stopped to eat lunch wherever they were, to resume their work and return to headquarters at 14:00. These changes were well received by the workers, as they could store the lunchboxes at headquarters, where they ate and rested, and did not have to return to work at the time of strongest sunlight.

One of the main problems under the new schedule was the lack of access to toilets, especially for the sweepers who worked in residential areas. Other problems mentioned were distance to worksites, misunderstanding of the task distribution (map), and difficulties related to changes of partners.

12.3 Ethnographic Data Collection

Ethnographic data collection before CL sessions consisted of individual interviews, collective work analysis (CWA), and observation along the workday. Thus the intervention agents became acquainted with the target activity and understood some of the problems associated with it.

The data showed that the workers rated their work tiresome; nevertheless, they often felt satisfied with the results at the end of day. They observed that the area to be swept had increased in in recent years (together with the expansion of the municipal urban area following the creation of new neighborhoods). They also mentioned some difficulty to understand the distribution of tasks (the sweeping map was continuously modified) and the resource to "backup" employees. Other aspects addressed were the fragility of materials and problems with meals (which had been solved before the beginning of sessions together with the changes in the working day) and toilet use.

As a result, the intervention agents were able to predict problems that could arise in the stage of questioning and used videos and the workers' narratives for double stimulation in sessions.

12.4 Change Laboratory Sessions

The formative intervention developed through meetings held at the workplace, namely, one room at headquarters. We held 15 sessions with the workers and further 5 with a group of consultants, which included company managers, CEREST representatives, and intervention agents.

Analysis and planning of sessions followed the cycle of expansive learning, beginning by questioning, then moving to historical and actual-empirical analysis to conclude in modeling of the causes of problems and solutions.

12.4.1 Questioning

While the CL sessions began in March 2017, there was a gap between the first and the second due to changes in the organization of work, including modification of the working hours and reduction of the number of workers.

The aims of the first session were to strengthen the relationship with the workers and explain to them the CL method in a simple and didactic manner. In this particular case, it was not possible to plan extra-session tasks, partly because some participants had a low educational level and partly due to lack of spare time to meet outside the workplace. Therefore, all the activities were performed during sessions using illustrative materials.

In the first session, we used the CEREST demand about the health of public cleaning workers as first stimulus. The second stimulus was a narrative of one of the workers who had participated in CWA. Double stimulation triggered a discussion of problems inherent to the sweeping activity, such as size of areas to be swept; frequency of sweeping; need for tree pruning and collection of asphalt pavement repair debris, branches, and grass; waste left by the homeless; lack of population awareness of the need to pick up dog stools; maintenance of public litter bins; and trash disposal hours, mainly among bar and restaurant owners.

The aim of the second session was to insist on the relevance of the study, as well as of the active participation of the workers, and to resume questioning in regard to the problem situation. For this purpose, as first stimulus we used a video made during the observation of a pair of sweepers along a working day, while the second stimulus was the sweeping activity system (AS). In this session, the participants talked about differences in the work performed in different locations, the brooms they made themselves, and adjustments to facilitate and speed up work. The group was able to visualize all the aspects involved in sweeping in the AS triangle (see Fig. 12.1) and finally made a list of the easy and hard aspects of work (see Table 12.1).

In the third session, the table of easy and hard aspects was used as first stimulus for questioning, and the second stimulus was some CWA statements. The participants discussed, first in subgroups and then all together, each listed aspect, such as changes in the working hours, excessive use of mobile phones by some workers, and removal of heavy waste, uniforms, and map drawing, among others.

All the problems and difficulties discussed in the first three sessions were used as first stimulus in the fourth. Images were used to represent the problems; the participants were requested to classify each image per individual and/or community responsible for a possible solution (see Fig. 12.2).

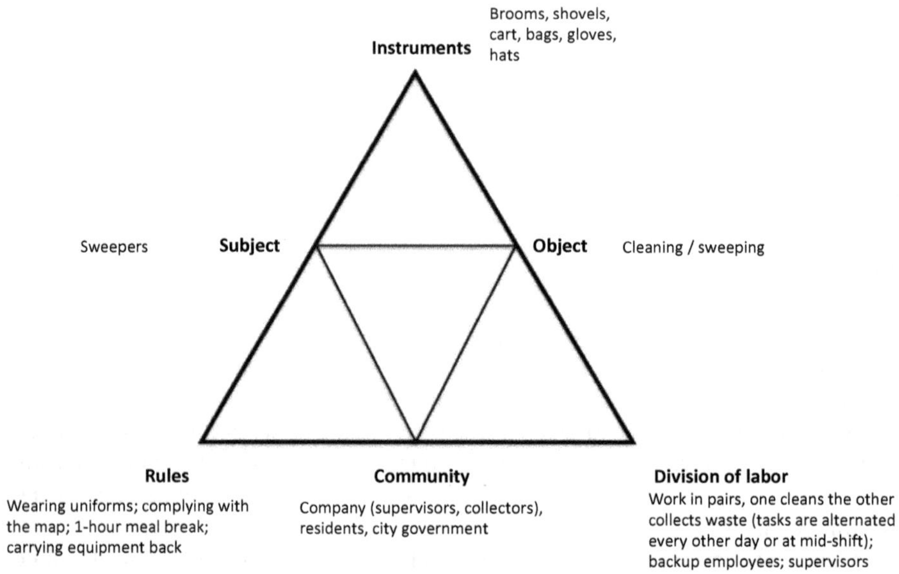

Fig. 12.1 Activity system developed by the group of workers in the second CL session. (Adapted from Engeström 2015)

Table 12.1 Easy and hard aspects of the sweeping activity according to the participants' narratives

Advantages/easy aspects	Disadvantages/hard aspects
Picking garbage with the feet	Rain (the street becomes very dirty the following day)
Three-member crews (including backup)	Partner does not go to work
Large brooms (homemade)	Unempathetic supervisors (some cases)
Breakfast	Residents' complaints
Work in pairs (not being alone)	Dog stools
Sweeping is nice	Homeless
Outdoor work	Sunlight
Early end of workday (free time for afternoon work)	Dust
Health insurance/pharmacy	Partner smokes
Meal voucher	Partner's use of mobile phone
Meeting new people	Partner wears headphones
Rain (not going out to sweep)	
Holidays	

Fig. 12.2 Organization of difficulties in the sweeping activity. Caption: population branches/grass, coworkers, supervisor, company, city government

During the discussions of changes, we moved from questioning to historical analysis, using a timeline as second stimulus. The participants talked about the successive companies which had previously provided urban cleaning services in the city and described the activity, materials used, and changes which had took place over time.

12.4.2 Historical Analysis

In the fifth session, which aims to discuss changes in the sweeping activity and ongoing contradictions relative to such changes, we used the timeline data to build a matrix of changes as first stimulus (see Fig. 12.3). The participants pasted images on a wallboard representing the AS (object, subject, instruments, rules, community, and division of labor) of each phase. The AS was used as second stimulus to investigate the contradictions in each period which had led to changes in the activity (see Fig. 12.4).

Double Stimulation
In the first two CL sessions (questioning), we (the intervention agents) perceived the participants had some difficulty to perform the suggested tasks due to their low educational level. We thus realized it would not be possible to ask them to perform extra-session tasks and that we needed to make optimal use of the in-person sessions. However, the participants also had difficulty with the tasks suggested for sessions, and some could not participate at all for being illiterate.

Therefore, we decided to use other materials for double stimulation to make discussions more productive. Images and videos were used as first or second stimuli, in addition to the figures and clippings used by the participants to illustrate their discussion of the presented stimuli. This methodological adjustment modified the pattern of participation, which became more effective and active at each subsequent session.

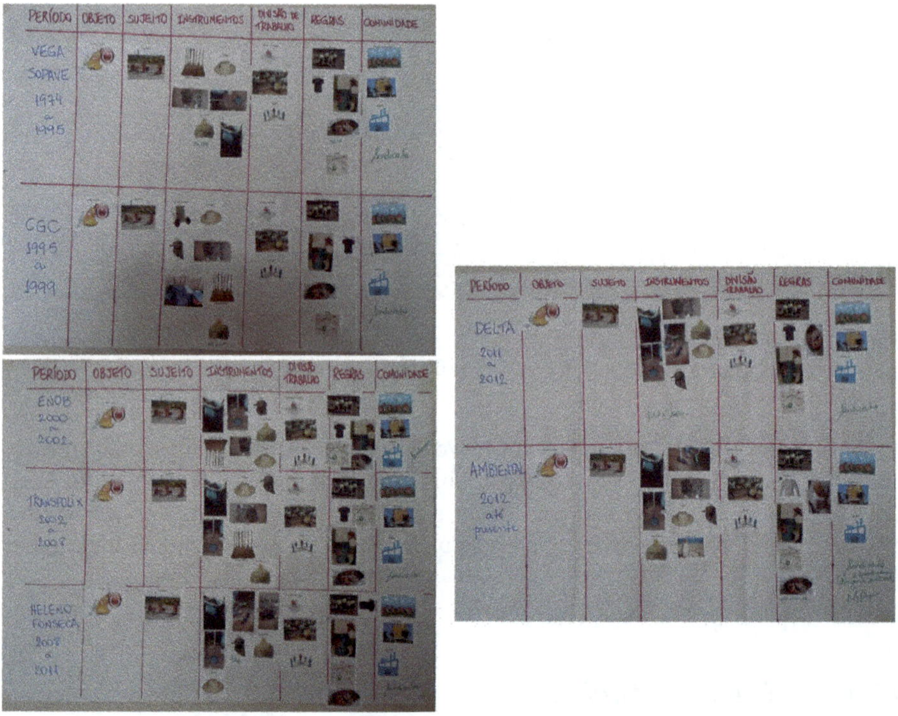

Fig. 12.3 Matrix of changes developed by the participants with illustrative materials. Caption: period, object, subject, instruments, division of labor, rules, community

The fact that the workers were not familiar with reading and writing did not hinder the development of the CL sessions, as the intervention agents found other means of expression to ensure satisfactory conditions for dialogue. As Paulo Freire observed in his works, one needs to look for situations which make sense to the involved group. This observation accounts for the use of images to represent the keywords to which Freire had resource to teach literacy to adults (Freire 1987). In the case of the sweepers, using images and illustrations as language instrument and record was a quite original solution, which paves the road for the application of the CL method to other groups, like immigrants, who do not, or partially, master the language, vocabulary, or accent of their new country.

Table 12.2 summarizes the process of double stimulation applied in sessions and the outcomes of the discussions based on illustrative materials prepared by the participants.

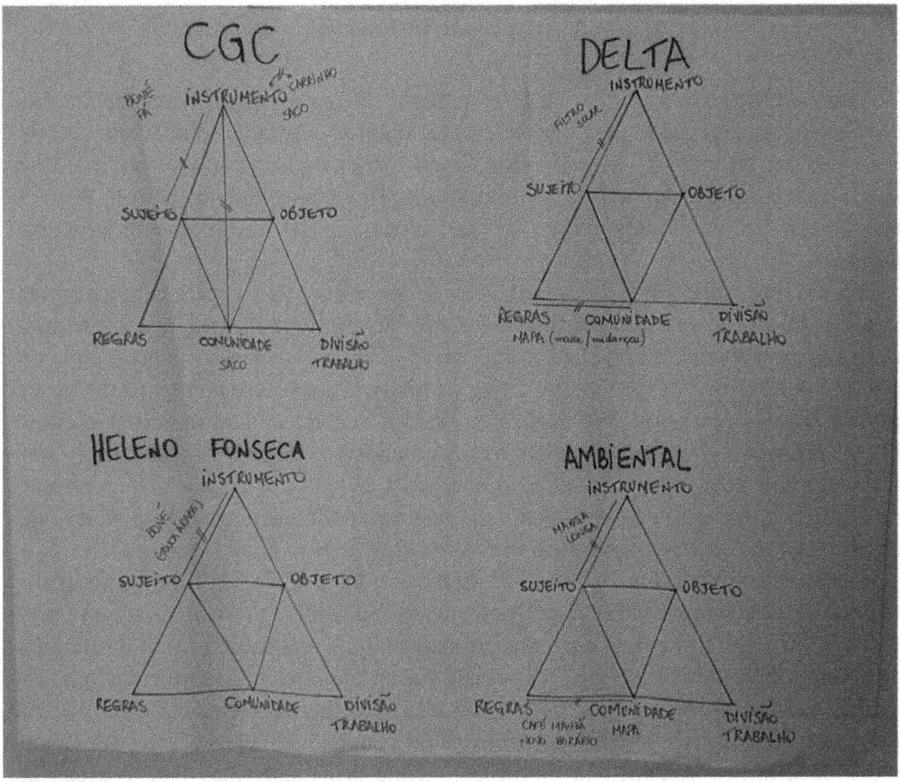

Fig. 12.4 Representation of historical contradictions in the activity system of each company. (Adapted from Engeström 2015)

Table 12.2 Summary and outcomes of double stimulation

Sessions	First stimulus	Second stimulus	Results
Session 2	Images representing workday situations	AS development (see Fig. 12.1)	Table of easy and hard aspects (participants' narratives) (see Table 12.1)
Session 3	Table of easy and hard aspects	CWA statements (ethnographic data)	Organization of the table according to perceived responsibility (see Fig. 12.2)
Session 4	Matrix of changes developed by the participants with illustrative materials (see Fig. 12.3)	Timeline	Timeline development
Session 5	Timeline and matrix of changes (see Fig. 12.3)	AS (see Fig. 12.1)	Analysis of historical contradictions (see Fig. 12.4)

12.4.3 Analysis of the Current Situation

The aims of the sixth session were to resume the discussion of the historical internal contradictions and start a new one on current internal contradictions. As first stimulus, we used the activity systems developed for each period in the timeline until reaching the current AS and its contradictions. To enhance the discussion, we used the table of problems prepared by the participants in the fourth session as second stimulus.

We noticed a quaternary contradiction between the AS of the city government and the AS of the sweeping activity. The city government had requested the company to cut costs, which resulted in reduction of the number of sweepers, and consequent redesigning of the sweeping map, which did not please neither the sweepers nor the local population. We further noticed a secondary contradiction between object and rules, since the sweepers were demanded to clean well and comply with the sweeping map, which included dirtier sites. Finally, there was also a primary contradiction within the object, as the sweepers were demanded to clean well (quality) but fast in order to accomplish the established metric goals (quantity).

In the seventh session, we used as first stimulus a table with the participants' suggestions of changes for each problem and asked them to think about their respective priority. The following were the priorities which emerged from the initial discussion: population awareness, restaurant waste, map design, shards of glass, and asphalt pavement repair debris. Next we analyzed the matrix of changes (second stimulus) to establish whether the object had changed somehow. The participants did indeed notice a change in the object: while previously they had to clean the downtown only, now the target was the entire city. This was the trigger for continuous redrawing of the sweeping map, which also led to complaints from the town population (as the frequency of the service had decreased in some neighborhoods, some residents were angry and complained to the sweepers). The participants also mentioned differences in street sweeping downtown and in residential neighborhoods and pointed out specific issues at each location. The result was a picture of differences between street sweeping downtown and in neighborhoods.

We held a session with supervisors, the company safety technician, and safety engineer to establish whether their opinions were the same as the workers'. The first stimulus was the presentation of the workers' narratives. The group analyzed them and agreed with the difficulties described. The second stimulus was the downtown vs. neighborhoods table, and we asked whether also they perceived differences. The group agreed with the described differences and explained that in neighborhoods the area to sweep is somewhat larger, because there are less obstacles to sweeping compared to downtown (e.g., parked vehicles).

In the eighth session with the workers, we returned to the downtown vs. neighborhoods table and asked them to try to identify common points. They all said that the problem of lack of population awareness was universal and that there was lack of information on proper waste disposal available to residents and store owners. As second stimulus we used a definition of term "awareness." The participants agreed that this was exactly what was needed, i.e., to provide information to the population to change its habits. As means they suggested flyers and communication through the media—radio, the Internet, newspapers, ads on wheels, and flyers pasted on their carts and public litter bins.

The Invisibility of Work
In addition to the aforementioned, one further contradiction impregnated all the relationships involving the sweepers, for instance, with supervisors and the local population. This contradiction concerned public invisibility. Despite the relevance of the sweepers' work to maintain cities clean, this population of workers seems to psychosocially vanish amid other people (Costa 2008).

Such public invisibility is the expression of two psychosocial phenomena, social humiliation and reification, which are latent in capitalist societies (Costa 2008). It is the result of a historical process marked by the social division of labor, which encompasses issues related to gender, ethnicity, and age. It points to both symbolic and material violence, in which context the relationships between workers and others—supervisors and the community—are weakened and have an exclusively economic nature, i.e., mere delivery of services (Costa 2008) which are only noticed when they are not provided.

During the first CL sessions (questioning stage), several difficulties inherent to the sweeping activity were discussed but also advantages and reasons for satisfaction. The participants succeeded in listing easy and hard aspects and also designed an activity system. Contrariwise, the local population systematically earned negative remarks as a function of their lack of collaboration, either due to lack of information or to the belief that some activities belong to the job description of sweepers. Some participants even stated that many residents are not even aware they sweep the streets where they live.

In the historical and actual analysis of the working conditions, the participants discussed priorities, in which context most of them pointed to changes in the sweeping map and the need to raise the population awareness of their activities and appropriate waste disposal.

The participants realized that a part of their difficulties concerned the local population, which needed to receive information in order to change. The participants discussed new solutions specifically aiming at raising the public awareness. Other topics did not only include the attention to waste disposal the population should pay to and the need to inform it on specific aspects of the sweepers' work but also the discrimination they suffer while doing their job. Some participants reported to have been victims of prejudice manifested in how they are treated by the local population, which associates them to "dirty" work, or simply ignores them, thus reinforcing the invisibility of this activity.

12.4.4 Modeling New Solutions

In the ninth session, we reviewed all the information the participants wanted to include in the flyers. We showed a flyer used in other cities (first stimulus) and asked what impact phrases would be necessary and whether they would like to create a character which could "talk" to the population. We conducted two separate discussion groups and used drawings to encourage ideas for the flyer (second stimulus). Then we returned to the general discussion to present the results of the group discussion and reached a consensus on the character and phrases to draw the readers' attention to the flyer. At the end of the session, the participants discussed who should help in the making and distribution of flyers, and most of them understood the need for a partnership between the company and the city government.

A session was then held with the group of consultants (manager, safety technician, safety engineer, a representative of the company communication department, and a CEREST representative) to discuss the aspects pointed out by the participants, mainly the ones which had been given high priority. This group agreed that the population did indeed lack information and that there were various means for divulgation. The group divided the tasks needed to make the flyers and survey the most problematic sources of waste, such as bars, restaurants, and snack bars. They also asked for more information on the problems related to the sweeping map. The intervention agents indicated the need to hold some sessions together with the sweepers exclusively allocated to residential neighborhoods, which suggestion was approved.

Three further sessions were then conducted, one to discuss the following CL steps and the other two at another company headquarters with sweepers allocated to neighborhoods. In these two sessions, the sweeping map and its changes over time were discussed, as well as the impact of such changes on the workers' performance and relationship with the local residents. As the frequency of sweeping had decreased in some areas and thus dirt and fallen leaves accumulated, the workload had become heavier, and the sweepers had to work faster to meet their goals. Changes in the frequency, and sometimes also in composition of the sweepers' crews, led to complaints from residents who did not understand the sweeping activity and even ignored that sweeping sidewalks was their own duty. The workers expressed dissatisfaction with having to sweep the curb of the street only, because the effect was not evident, as most residents did not sweep the sidewalk.

A meeting was then held with the group of consultants to discuss the flyers' content and map drawing. It was established that the current map would not be changed for the time being. The group also approached the difficulty and ineffectiveness of sweeping the curb of the street only. At this time, the downtown sweepers' survey of problematic waste sources had not yet been concluded.

After a time needed for the flyers artwork, we held two meetings with the group of consultants to analyze them and one further session with the group of workers to approve the flyers' design and content. The two flyers thus developed, one by the cleaning company and one by CEREST, were approved and matched exactly the content developed by the sweepers. The latter expressed satisfaction with CL, because they felt their voice had been heard for the very first time.

The aim of the following session was to discuss means of divulgation of the informative material and was very productive. In addition to suggesting means, such as social networks and printed flyers, the participants also considered divulgation at schools to raise the youths' awareness on the sweeping activity and on how to dispose waste appropriately, thus promoting an educational action targeting the new generations. We also discussed making short videos and documentaries. In this regard, parts of a documentary were used as first stimulus and narratives from sweepers in other cities (newspaper articles) as second stimulus. The participants also discussed prejudice and how it would be overcome, where the population needs to understand better their work.

A final session was held to discuss the results achieved along the CL. We used a video made by a local store owner on the sweeping activity. We discussed aspects included in the video, such as lack of population education on public cleaning and lack of recognition of the sweepers' work. The group agreed that there were not many means to educate the population and expressed their belief that their work in sessions would contribute to raise the awareness on everybody's duties and on the sweeping job. Once again they discussed prejudice and lack of recognition and how the suggested videos could help make their work and its effects—i.e., a clean city—become more visible.

Therefore, this CL reached the stage of solution modeling and is now moving to the implementation of the ideas suggested in sessions. Some aspects, such as the sweeping map, might come to be discussed in the future, as the company decided not to change it over a long period of time. Nevertheless, the city government's demands might influence its design and distribution in the future.

The main point addressed in the present CL was the population awareness of the city cleanliness and the sweeping activity. The intervention agents will follow up the stage of implementation and help in the making of the workers' educational videos.

12.5 Resulting Innovations

As explained all along the present chapter, the intervention agents felt the need to make adjustments to the CL methodology, such as excluding extra-session tasks as a function of the low educational level of the participants. They were in a limit situation which led them to discover other possibilities:

> Ultimately, themes are the content and the container of limit situations; the tasks implied by them demand limit actions. When themes are hidden by limit situations, and thus are not clearly perceived, the corresponding tasks—this is, the responses of men through a historical action—cannot be performed in an authentic or critical manner. In such condition, men are unable to overcome limit situations and discover that an unexperienced possibility precedes—and is contradiction with—them. (Freire 2016, p.61)

Almost instinctively, the participants began to use other forms of language, such as figures and images, and the more these stimuli were presented, the greater their participation and agency in the CL process. In agreement with the CL methodology and Freire's ideas, we were looking for a method to serve as a tool in the encounter of workers and intervention agents:

> Hence our initial distrust of alphabets which intend to offer the arrangement of graphic signs, further reducing the alphabet to the status of object—rather than of subject—of literacy ... (Freire 2016, p. 78)

In both Vygotsky's psychology—which is the basis of CL—and Freire's pedagogy, work has an ontological nature. This is to say, "humankind is born and develops within each human being through work" (Poli 2007, p. 60). On these grounds, we seek to enhance the agency of workers.

All along the CL sessions—questioning and historical and actual-empirical analysis—the workers were able to appropriate their own work to culminate in the modeling of new solutions:

> [...] As 'beings-in-situation,' men are rooted in time and space conditions which mark them and are also marked by them.
>
> They will reflect on their own 'situationality' inasmuch as it challenges them to take a stance in its regard. *Men are what they are because they are in a situation* [original italics]. The more they reflect critically on their own existence, the more they will act on it, and more will be.
>
> Education and thematic research within a critical conception of education are just different moments of a same process. (Freire 2016, p. 66)

To conclude, an innovation specifically brought by the CL described here is that at the end of the sessions, all the participants' narratives were about being heard, in addition to succeeding in expressing their feelings related to social invisibility. The participants succeeded in acting on their work processes and felt they could play a relevant role in the development of improvements in regard to their job and its effects, which extend to society at large.

This study was financed by the Brazilian Labor Prosecution Office (MPT).

References

Bandeira, L. M., & Almeida, T. M. C. (2015). A dinâmica de desigualdades e interseccionalidades no trabalho de mulheres da limpeza pública urbana: O caso das garis. *Mediações-Revista de Ciências Sociais, 20*(2), 160–183. https://doi.org/10.5433/2176-6665.2015v20n2p160.

Chillida, M. S. P., & Cocco, M. I. M. (2004). Saúde do trabalhador e terceirização: Perfil de trabalhadores de serviço de limpeza hospitalar. *Revista Latino-Americana de Enfermagem, 12*(2), 271–276.

Costa, F. B. (2002). *Garis - Um estudo de psicologia sobre invisibilidade pública* [Dissertation]. São Paulo: Universidade de São Paulo.

Costa, F. B. (2008). *Moisés e Nilce: Retratos biográficos de dois garis: um estudo de psicologia social a partir de observação participante e entrevistas* [these]. São Paulo: Universidade de São Paulo.

Engeström, Y. (2015). *Learning by expanding: An activity-theoretical approach to developmental research*. New York: Cambridge Press.

Engeström, Y., Virkkunen, J., Helle, M., Pihlaja, J., & Poikela, R. (1996). The Change Laboratory as a tool for transforming work. *Lifelong Learning in Europe, 1*(2), 10–17.

Freire, P. (1987). *Pedagogia do oprimido*. Rio de Janeiro: Paz e Terra.

Freire, P. (2016). *Conscientização. Trad. Tiago José Risi Leme*. São Paulo: Cortez.

Gemma, S. F. B., Fuentes-Rojas, M., & Soares, M. J. B. (2017). Agentes de limpeza terceirizados: Entre o ressentimento e o reconhecimento. *Revista Brasileira de Saúde Ocupacional.*. [Internet], *42*, e4. https://doi.org/10.1590/2317-6369000006016.

Gomes, C. C., & Oliveira, R. S. (2013). Agentes de limpeza pública: Estudo sobre a relação prazer/sofrimento no ambiente laboral. *Psicologia: Ciência e Profissão, 33*(spe), 138–153.

Lopes, F. T., Maciel, A. A. D., Carrieri, A. P., Dias, D. S., & Murta, I. B. D. (2012). O significado do trabalho para os garis: Um estudo sobre representações sociais. *Perspectivas em Políticas Públicas, 5*(10), 41–69.

Monteiro-Cocco, M. I. (1997). *Reestruturação produtiva e o setor saúde: Trabalhadores de enfermagem em saúde coletiva* [these]. Ribeirão Preto: Universidade de São Paulo.

Pataro, S. M. S., & Fernandes, R. C. P. (2014). Heavy physical work and low back pain: The reality in urban cleaning. *Revista Brasileira de Epidemiologia, 17*(1), 17–30. https://doi.org/10.1590/1809-4503201400010003ENG.

Poli, S. M. A. (2007). Freire e Vigotski: O diálogo entre a pedagogia freireana e a psicologia histórico-cultural [These]. São Paulo: Universidade de São Paulo. https://doi.org/10.11606/T.48.2008.tde-16062008-133010.

Seligmann-Silva, E. (2011). *Trabalho e desgaste mental: O direito de ser dono de si mesmo*. São Paulo: Cortez.

SIEMACO - Sindicato dos Trabalhadores em Empresas de Prestação de Serviços de Asseio e Conservação e Limpeza Urbana de São Paulo. (2011). *Perfil dos trabalhadores em asseio e conservação e limpeza urbana de São Paulo*. São Paulo: SIEMACO.

Virkkunen, J., & Newnham, D. S. (2013). *The Change Laboratory: A tool for collaborative development of work and education*. Rotterdam: Sense Publishers.

Chapter 13
The Clash Between New and Old Models of Surveillance System: A Case Study of Change Laboratory in a Workers' Health Reference Center

Gislaine Cecília de Oliveira Cerveny, Marina Zambon Orpinelli Coluci, Renata Wey Berti Mendes, and Rodolfo Andrade de Gouveia Vilela

Abstract This chapter presents the results of Change Laboratory (CL) procedures in a Workers' Health Reference Center (CEREST) to improve working conditions and enable its staff to use CL methodology in their own territory in the future. One year after the 3-year intervention, the staff was interviewed to assess its results. The findings reveal that part of the actors expanded the work activity's object, promoting changes in work organization and using the methodology as a surveillance tool. However, there is always resistance to the new model, due to pressure for quantity over quality, lack of engagement of some people within the team and hesitations of the management. More effort and time will be necessary to consolidate the new model by the workers.

13.1 Introduction: Context of the Application of the Change Laboratory Method

Brazilian public health policies are programmes and actions carried out by the federal government in order to implement health services established by law: according to 1988 Brazilian Federal Constitution, "health is a right of all people and an obligation of the State". The largest public policy programme, regulated in 1990, is SUS, Brazilian Unified Health System (from the Portuguese Sistema Único de Saúde), which is a model of care based on epidemiology, social control, decentralization and regionalization with a municipal base.

G. C. O. Cerveny (✉) · M. Z. O. Coluci · R. W. B. Mendes · R. A. G. Vilela
Department of Environmental Health, School of Public Health,
University of São Paulo, São Paulo, SP, Brazil

R. A. G. Vilela et al. (eds.), *Collaborative Development for the Prevention of Occupational Accidents and Diseases*, https://doi.org/10.1007/978-3-030-24420-0_13

Its principles and guidelines are universality, integrality of assistance, equity, decentralization and community participation (Vieira da Silva and Almeida Filho 2009; Carvalho 2013).

In workers' health, SUS provides promotion and protection activities, as well as recovery and rehabilitation for those exposed to the risks and damages arising from working conditions (Sanchez et al. 2009).

National policies on occupational health and safety are implemented by a National Network for Integral Action on Workers' Health (RENAST in Portuguese) (Brasil 2006), implemented through Workers' Health Reference Centers (CEREST in Portuguese) (Dias and Hoefel 2005). These centers are responsible for technical support, continued education, coordinating assistance projects and occupational health surveillance in its region (Brasil 2006; Galdino et al. 2012).

Change Laboratory (CL) (Engeström et al. 1996; Virkkunen and Newnham 2013) application in a CEREST happened after a training course on this methodology with its authors, as an activity integrating "Work accident: from analysis to social construction of change", a research project supported by the São Paulo Research Foundation (FAPESP in Portuguese).

Considering that CEREST workers took part in the training course, the idea of using the institution as a pilot would have a double benefit. On the one hand, it would be a way of the team mastering the methodology (which could become a technical tool for intervention, contributing to the expansion of prevention actions developed by the group); on the other hand, it would offer the possibility of identifying and discussing CEREST's problems and difficulties to fulfil its duties.

This chapter aims to describe the expansive learning cycle experienced by the CEREST's participants through the development and consolidation of a new model of occupational health surveillance. Since the process did not have management's adherence, we will emphasize the challenges faced by the actors who led the learning process in a context of struggle between the new and the old model of surveillance.

13.2 Procedures

We started the process in May 2014 by presenting CL method to CEREST managers. After their approval, we used ethnographic research to collect data about the center's history, its operation and the main problems faced by the team. Semi-structured interviews with professionals who worked both in surveillance and in assistance activities[1] were conducted, as well as with coordinators of CEREST and Municipal Health Surveillance Area, institutional partners of the center.[1] Part of the questions had been previously applied to CEREST workers in 2012, during the first CL course in Brazil.

[1]What is your activity? What is the purpose of CEREST? How do you contribute to this goal? What are your experiences of success and failure? What contributed to success? What contributed to failure? Do you think CEREST has made any progress in its history?

We analysed CEREST's activities evolution and identified changes in quantity and quality of its demands. Ethnographic research findings were also used as mirror data, used to offer key information to participants in the discussion about activity contents over time—past, present and future.

In August 2014, all CEREST workers, including administrative, surveillance and assistance areas, were invited to a CL presentation. The only two requirements for CL sessions were voluntary participation and commitment to attend to sessions. Furthermore, all participants were aware that they could leave the sessions any time, without consequences.

According to CL method, all sessions were filmed to make future analysis easy and to guide the implementation of formative sessions. All recorded interviews and CL sessions were transcribed and participants' speeches analysed, in particular regarding three aspects: learning by the double stimulus, agency of the actors and their multi-voicedness.

Twenty-three sessions took place between September 2014 and April 2017, fourteen of which analysis and planning sessions and nine follow-up sessions. Most of surveillance and assistance workers accepted to participate in the intervention research, and the number of participants during sessions ranged from 9 to 14. Part of the group resisted participating on the problem mapping situation and did not come to these sessions. They returned afterwards. On the other hand, nineteen participants were present during the sessions to create the new model.

In July 2018 we returned to CEREST to interview the workers according to the Bardin content analysis method (Bardin 2009). Section 13.4 of this chapter presents interviews' results.

13.3 Description of Procedures and Results of the Sessions

During the initial sessions which occurred on September 10 and 17, participants were divided in groups and discussed what were the objectives of CEREST and whether they had been achieved or not. Most people talked about the goals set by the Ministry of Health and noted that they were not widely achieved due to lack of tools for promoting transformation. It was a consensus that a diagnosis could be made, but intervention was limited because "there weren't enough resources to achieve CEREST's goals". Another point raised by them was the lack of information and communication between areas, due to an increase in work demand which caused teams to carry out their activities in isolation.

In session 1, groups were given a task of answering the following: (a) What were CEREST's objectives? (b) How to achieve these goals? (c) Are the objectives being achieved? (d) What are the problems and how are they related to each other? (e) What are the causes of the problems?

In session 2, participants brought their answers and elaborated a table listing the main problems by sector and analysis of their causes. Surveillance sector problems were considered the most urgent, so they became CL focus.

In sessions 3 and 4, as an initial stimulus, the discussion on "What is CEREST's object?" was deepened, in particular to distinguish accident inspection from worker health surveillance. A second stimulus was the definition of surveillance: "Continuous and systematic action in order to detect, understand, research and analyse health problems determinant factors related to work processes, in order to plan and evaluate interventions to control or eliminate them".

Workers began to reflect on CEREST's activity and to voice different opinions about it. They began to seek explanations and ways of overcoming challenges in that activity and concluded that they were carrying out inspection activities and not surveillance. This conclusion left them discouraged and unmotivated by work.

Some participants had already taken an introductory course on CL in 2012. They were tasked to build a CEREST timeline. This material was brought in the CL sessions to initiate a historical analysis of the activity and to stimulate discussions about changes in their activities and its consequences.

Since its creation twenty years ago, CEREST has undergone numerous changes, especially regarding object and tools used in the activity. CEREST workers considered there was progress in working methods, especially regarding relationship with the community (companies, unions and workers), which brought population's recognition of their work and increased both the demand for work and denunciations.

We used another stimulus with the timeline: marking in its tasks quantities in different periods. The number of inspections increased from 15 in 1997 to 152 in 2008 and 387 in 2013. Task diversity also increased. The main tasks in 1997 were inspections and meetings in negotiation tables, whereas in 2006 the main tasks were elaboration of infraction notices, investigation reports of serious and fatal accidents, notifications to companies and participation in public hearings. Although the number of workers also increased (from 7 in 1997 to 16 in 2013), it was not enough to follow the increase in demand. In CL sessions, all participants were unhappy with work overload. They pointed out that pressure for quantity was the main factor for their actions' low quality.

13.3.1 Activity System's Contradictions Analysis

The main contradiction found in CL sessions was the dilemma between quantity and quality, as can be seen below:

> (...) we evolved, but we have to study more, since the demand is large and in service, we have to give answers to the people who knock on our door, there is little or no time to study. And CEREST has become much more well known, so there is more demand now than three years ago, a lot of demand and a lot of work. (Subject A)
>
> There used to be more teamwork, everyone learned more, and successful results were commoner. (Subject B)

Another contradiction was found between instruments and rules, as shown in Subject C's speech on the difficulty of applying in-depth analysis tools due to high demand:

> I apply the MAPA (Analysis and Prevention of Accidents Model) tool, but I think it's difficult; it takes time to do it. As it takes time and demand is high because of new accidents, it cannot be applied as we would like. (Subject C)

In order to deepen the analysis of the contradictions, we used the activity system triangle as a stimulus. With this figure, participants were invited to identify specific elements of the surveillance activity system and the existing tensions between these elements in daily work. Initially they chose to work on a case they deemed unsuccessful; in the following session, with successful ones. It was possible to find primary and secondary contradictions between all elements of the activity system, namely, subject, instrument, object, rules, tasks division and community. When comparing success and failure, participants identified key factors to increase the chances of improving accident prevention: a collective subject, tools that allow in-depth analysis, enlarged community with the participation of unions and especially a work of surveillance built over the years with the company. Another relevant fact raised by them was the decrease in accident rates and, more significantly, in their severity, especially in cases where surveillance activity in the company occurred in a more negotiated way. In companies where the activities were restricted to inspections, neither reduction of accident rates nor of their severity was observed.

13.3.2 Modelling a New Solution

Participants presented a new design of the activity system (see Fig. 13.1). The subject, initially an individual (an inspector—technical or safety engineer), became collective with the participation of professionals from the care sector (medical doctor, sociologist and nurse). The object, which was inspection, became surveillance. The instruments included team and negotiation meetings. The existing rules—answering to complaints—became planning, consensus autonomy (worker autonomy, the actions based on a consensus of the team) and participation in team meetings. Tasks, which were previously classified in assistance and inspection sectors, became divided in emergency actions, planned actions and information room.

The new model based on changes in CEREST activity system was revised, and participants reached a consensus on how solutions discussed in analysis and planning sessions should be tested in the service. They agreed that at least 4 months would be necessary to implement work processes' proposed changes.

13.3.3 Follow-Up Sessions

The new model implementation lasted for 6 months, and, in November 2015, the follow-up sessions began. Prior to these sessions, we interviewed three workers and CEREST's coordinator to generate mirror data. We verified that there were several difficulties with the new model: some workers were still afraid to put their opinions forward; teams' responses to emergency actions relied on participants' experience

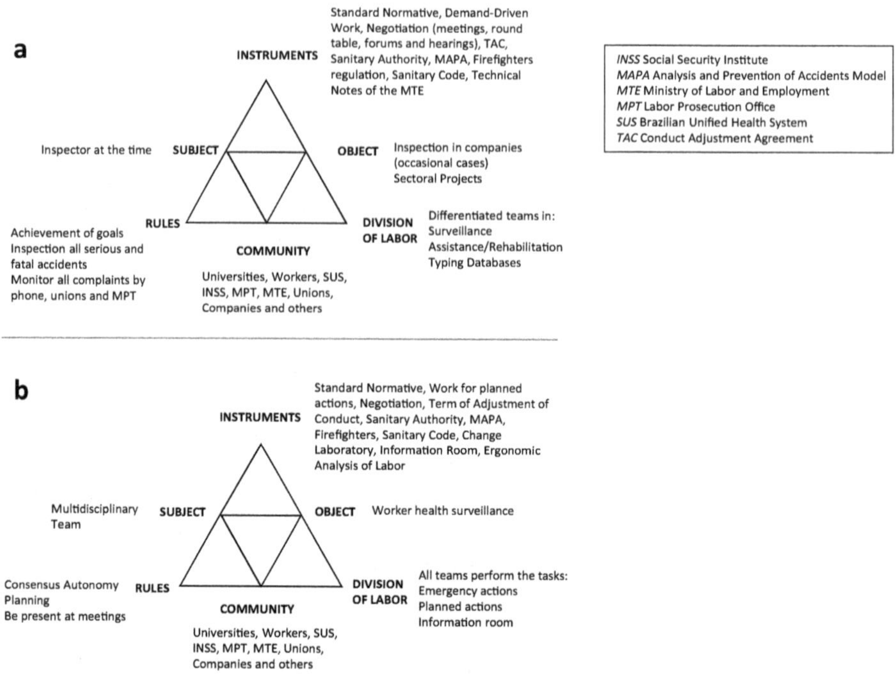

Fig. 13.1 (**a**) CEREST activity system before CL, (**b**) CEREST activity system after CL. (Adapted from Engeström 2015)

with certain tasks to be effective; and some areas evolved more than others, with the information room (IR) (see Fig. 13.1) being the least developed due to limited staffing and lack of understanding of its objectives by members of the group. Despite the difficulties, all interviewees pointed out improvements in communication and integration between workers, who were able to exchange learnings after CL sessions. Another important achievement was the creation of a Permanent Education Nucleus (PEN), which came to be a new area inside the CEREST, and the continuity of weekly meetings to discuss their actions, even in the absence of interventionists.

Three initial CL follow-up sessions were carried out to evaluate the model and to identify possible issues to be corrected. In session 1, subjects spoke about the difficulty of trying and working from a new perspective, while still under pressure for high demand, without being able to organize and prioritize their activities. As positive aspects, they identified a change in perspective (new conception of surveillance), the persistence of the team meetings, teamwork, the information room (IR) structuring, the emergence of PEN and the spread of knowledge to all within the service, mainly in relation to instruments used. As negative aspects, they pointed out that some workers in care sector were not engaged (they did not show interest in the discussions and were unaware of the changes in the service); the teams' composition, with differences in workers' attitudes, resulted in more work overload to some of them; and there was a lack of clarity as to emergency actions and how to work with them. They also pointed out the need to qualify emergency and planned actions, to develop IR and to integrate PEN into CEREST's services as a whole (see Subject D speech). The main discussion in session 1 was about whether all actions could actually be planned. At the end of this session, participants proposed the following task: to choose actual cases of success and failure during the implementation period and describe them to one of the interventionists:

> ... We need to redefine IR team to achieve a better labor division in it, and plan actions of teams thinking about affinities. (Subject D)

A fall accident was chosen as the first stimulus to begin the reflection on the process in the second follow-up session. A flowchart presented a case's evolution up to its conclusion, which involved other communities (companies, unions and workers). This mirror data was the basis for a discussion about differences in working practices that involved several people. It helped in producing more ideas and engaging all participants during team meetings. Participants identified that dialogue within the teams changed accident analysis, broadening discussion about work organization and ways to prevent fall accidents. Finally, a second stimulus that involved a two-dimensional model helped in expanding the analysis from the work made individually (an inspector at the time) to teamwork and from the local action to surveillance and prevention actions.

After using the first and second stimuli, we asked participants: "What is missing? How can you develop CEREST's organization to get to teamwork for surveillance and prevention more effective"? Discussions that followed pointed to the need for further action, such as creating a space to discuss identified contradictions and to prioritize cases for in-depth analysis and concentrating all demands on the information room. At that point in time, a conflict came up between CEREST's staff and its coordinator. Staff proposed that the coordinator should not take part in these group sessions. She agreed to it, arranging to meet individually the interventionist to address management issues. At the end of this session, all members agreed that the service should be organized into two areas: IR, responsible to analyse all the demands and to distribute tasks to the teams according to priority (Subject E), and PEN, responsible for preparing courses and lectures and disseminating information on innovations, among other tasks:

> We believe that IR should provide quality data. For this, we have first to lay a foundation, then we improve on quality. (Subject E)

Participants came up to the third follow-up session with objectives for each area, how they should work to achieve them, the necessary instruments for that and how the areas should be structured. They created a timetable for implementing the new model—a task proposed at the end of the previous session. They discussed IR and PEN activity systems, the division of labour and about resources from other communities that could be used to help them, especially regarding IR database. At the end, a schedule for IR and PEN development and organization was set to be implemented over a period of 4 months.

The fourth follow-up session occurred in April 2016 and began with a summary of IR organization implementation. There had been difficulties, and the group discussed the need for some staff members dedicating full time to data analysis. They also discussed the demand flow, which had improved since it had gone to IR. There a demand was analysed, classified by type of action that should be taken and how many people would be involved, without fixed teams. During the discussion, a conflict arose about the incoming demand and how it influences the distribution of tasks, which led them to questioning how to use indicators to signal priorities. All participants said that the old system resisted changes, mainly through the coordinator, who was under pressure from higher management. In addition, they had not been able to establish partnerships with other institutions. The suggested task that resulted from the session was to continue to improve IR database and to organize PEN activity system.

In the fifth session, participants presented their tasks but reported difficulty on the division of labour. To stimulate discussion, the interventionist brought some surveillance concepts, and the participants talked about CEREST's object again, which they defined as extended surveillance. Based on those concepts, the interventionist presented the triangle of the activity system again and asked if they could complete it considering CEREST as a whole, integrating IR and NPE. This should facilitate with services organization of and with division of labour. They discussed during the session and received a task of presenting the activity system in the next session.

The sixth session began with the presentation of the group's activity system. Participants concluded that their activities should be triggered by the IR but developed by the assistance and sanitary inspection areas and PEN. They then discussed about difficulties and strengths of that model and suggestions for working with it. One of the suggestions was to improve communication, since it occurred that, when staff went to meetings or external events, it was difficult to share information about it with the group due to lack of time and space. All subjects agreed that individual demand had decreased, but some difficulties remained. For instance, weekly general meetings focused more on administrative matters rather than case studies and the lack of clear priorities, of strategic planning and of the establishment of new rules and creation of partnerships. At the end of the section, it was proposed to draft CEREST's internal rules systematization.

A proposition for the text of internal rules was presented at the seventh session, which brought differences in formatting and description of the sectors. In group, it was decided on the document's format and how each area should be described—introduction, objective, attributions, criteria, strategies, goals and actions. The second part of the session was dedicated to discussing the assistance sector and the relationship between CEREST and the Municipal Health Network. Participants agreed that some changes were necessary, but some of the topics discussed would increase the demand for the center and even change its purpose. At the end of the discussion, IR and PEN staff detected the need to construct epidemiological criteria by severity of accidents in order to prioritize their actions.

At the eighth follow-up session, participants presented better structured internal rules and discussed how to record data from a specific occurrence. These data would include information about the worker, the company, patient care actions, the relationship between illness and work activities and the assignment of new cases to the teams. They realized that it was necessary to carry out strategic planning at that time and agreed that PEN would be in charge of inviting a specialist to help with it. Finally, they were divided into three teams to discuss and write down criteria for extended surveillance based on low, medium and high complexity. At the end of the session, it was proposed to complete standard writing, including criteria and CEREST actions. After the session, the group undertook a training on strategic planning, and CEREST moved to a new building.

The ninth and final follow-up session was held in April 2017, when participants summarized the changes that occurred since the previous session. CEREST staff had decreased, and the internship programme was cancelled due to a municipal budget cut. At the same time, participants began the strategic planning with the construction of the flowcharts of CEREST actions. New ideas emerged from this activity, such as seeking partnerships with other CERESTs and related communities and information integration at the Sanitary Surveillance Information System. They also suggested a change in the way denunciations were received. Complaints, which were received through different mediums such as telephone and cell phone text messages, should now come through the city hall website only. A member of staff would be responsible by accessing it daily. Complaints received from the website would then be checked and replied to by CEREST after any action was carried out. Staff also began to meet weekly to discuss and plan technical actions.

13.4 Consolidation of New Practices

After nine CL follow-up sessions, participants were left with the task of consolidating these new practices before involving the community represented in the activity system. One year later, we decided to evaluate the entire CL process in CEREST, which had begun almost 4 years before. Interviews with semi-structured, guiding questions were then carried out with 12 CEREST workers.[2] The answers were analysed according to the Bardin content analysis method (2009) and separated into the following five categories:

[2] How did CEREST change after the end of CL sessions? Why were the meetings interrupted? How is CEREST now? How is the division of labour? Did the change take place in the same way in all sectors? How is the relationship with management? What if the CL was not used? Is it worse or better than it was before? Is there any change that remained? In relation to the change of personnel, did the people who entered have any training regarding the new model?

13.4.1 Improvements in Work

Interviewees thought that, after CL sessions, CEREST activities were expanded. Initially, activities were focused on inspection actions and then on planned, concerted interventions made in companies of the same sector (the so-called promo, e.g., *promo* of the electric sector work and *promo* of street cleaning work).

There was also an expansion of the covered community that was expected with the collective interventions:

> We started looking for support elsewhere. This was the expansion of the community, because we are closer to other CERESTs, to the Labor Prosecution Office (MPT in Portuguese), so there was expansion of the partners.

The fundamental change pointed by staff regarding the new activity configuration was the information room creation. It provided more credibility to CEREST among the companies in which they operate:

> As data are more organized in the information room, CEREST gains credibility, because when a company has an increase in the number of accidents, this is detected and CEREST communicates it, so that the company knows that CEREST is watching.

It also provided activities' integration and facilitated data processing, which came from the tools they already had:

> Work became more integrated, more united (…) all accidents, even if we do not manage to inspect, are better documented and everything is archived and organized.

The interviewees considered that their initial objective—learning about CL tool and its application—was achieved:

> Learning was a giant leap in personal terms. Things that used to take a lot of time to do, using CL concepts and tools were more feasible.

Although with some difficulties, many have tried to apply CL concepts in surveillance actions in companies and in *promos*:

> It is difficult to do the CL in the company, but we have used its concepts and instruments in the negotiation process. Looking at the rules, instruments, etc., for diagnosis was very important.

They recognized that they already knew how to make good diagnosis before, but intervention in companies turned out to be more focused:

> Today we are able to move forward in a longer intervention process. CL was important to the teams' maturity; we use teaching during interventions. In one company, we used a training intervention, through a cooperation agreement; the goal was always to discuss, without the company's legal sector participation, with support from the MPT and it was delightful.

13.4.2 Criticism of the Method and Its Application

Although the interviewees evaluated the expansive learning process with CL method as positive, they also expressed criticisms and unsolved situations. They criticized the way interventionists (called by them "academics", because they understood that the intervention occurred in a context of academic research) led the phases:

> The university failed in some ways. There were interviews that were scheduled and didn't happen. The team failed to keep up with individual interviews in order to figure out possible improvements, so there was a gap.

They also considered that at the end of the intervention, some contradictions still remained; nevertheless, they trusted the tool as one that could help them to overcome those contradictions:

> It is not a tool that persisted, that was absorbed by the service ... when the CL meetings ended, we were a little lost and without stimulus to take the next step. There was a gap ... I believe we need some strategy to resume meetings here. The service will create more noise, so maybe resuming CL is good.

13.4.3 Improvements that Did Not Persisted After CL Sessions

The activity designed during the interventions provided a set of collective actions planned during team meetings but that planning stopped when CL process came to an end:

> Due to the fact that team meetings stopped, we do not have a collective planning anymore.

Without team meetings, work was once again carried out individually, jeopardizing gained achievements and allowing the resurgence of the quality versus quantity contradiction:

> The biggest loss regarding the CL was that teamwork used to intervene rolled back and work was again done individually. This is a step back, along with the idea of inspection.

13.4.4 Remaining Problems After CL Sessions

According to the interviewees, staff members' absence in sessions hindered the process of expansive learning and activity redesigning:

> At the end of CL cycle, not everyone engaged in the process, not everyone was present at all sessions, so that was a hindrance. People who did not participate in those moments became resistant to the process.

The conflict with management also persisted, generating some difficulties. Subjects criticized the lack of leadership and strategic direction of actions of managers:

> The lack of managerial work, for instance creating a timetable, a schedule of actions and how far they should go, is very important, because then technicians lack direction and do whatever they think it is right (…) An important factor is that the coordinator can change what was decided during CL; he tries to go back to the old system and resists to changes (...) The meetings were useless. In addition, because of the coordinator's behaviour, people ended up not wanting to participate anymore.

The lack of engagement of part of the staff was also evaluated as having a negative impact on changes:

> At Friday's technical meetings, most technicians were absent. There was no interest in changing that reality, in receiving more information.

There were also community issues perceived as difficulties:

> There is staff shortage also in federal and state spheres to address these issues ... Many of CL issues have not yet been solved. There are contradictions. The concept "victim's guilt" ... this is very strong, so you cannot deconstruct it ... the problem of CEREST, like every company, is that it is at mercy of strong external forces that affect the entire system.

13.4.5 New Problems that Arose After Applying CL Method

Regarding difficulties that CEREST still has to face, interviewees mentioned new problems that arose after the end of CL sessions. The service lost several key staff:

> Six people were fired, two retired. Five were hired (one will retire next year), but none has knowledge in ergonomics. Instead, their approach blamed the victims of unsafe behaviour ... There was significant knowledge loss ... part of the staff was overwhelmed by the layoffs.

CEREST's new facilities location was also presented a problem that interfered with the service:

> It is far from the bus terminal. We are not sure, but it can be for this reason, anyway there was a great reduction in users demand.

13.5 Final Considerations

A new expanded object of surveillance (new model) has been consolidated through the CL. Daily practice disorders analysis, a detailed study of what does really work, don't work, historical analysis and other learning devices had shown that the complex object of surveillance calls for new instruments that can deal with this complexity such as the activity-centred ergonomics, the MAPA, etc. The information room that verifies the priorities is important to enable the high demand received by the small team and what kind of action they should take. The complexity of the object also requires for a new work division (teamwork X the individual work) and for more democratic and horizontal relationships, creating a more flexible and creative environment.

Activity system analysis has shown they need more time and to expand their community, their partners. These challenges of the new model fight with the old model that resists in the form of incompatibilities in all elements of the system (tertiary contradictions). The new model also clashes with neighbouring surveillance systems (Labor Prosecution Office, Ministry of Labor and Employment, sanitary surveillance, health secretary, etc.).

The case also reveals that the object's primary contradictions (quantity X quality)–or division of labour contradictions (autonomy and cooperation among staff X rigid hierarchy)–generate continuing to tension over activity and produce new and old disorders that make it impossible for the workers to act. Although CL started upon agreement with the manager, she later became contrary to the changes proposed during the sessions, which pushed part of the staff to leave the meetings.

This conflict between the new and old models has been experienced and faced by the team that tried to consolidate the new method of work. However, this effort did not have the support of management, which made the process weak and stressful for those who lead the transformations of the object. A very long implementation period caused a certain disruption in the discussions. When the sessions resumed, part of the staff was unmotivated due to unsatisfactory results from previous steps. Changes made during follow-up sessions optimized some services and provided new ideas. However, again, at the end of the process, participants couldn't keep up with neither the expansive learning cycle nor the necessary periodic meetings, since the manager didn't want it. We believe that these problems could be reduced by investing in a learning process with the management—a challenge for new CL experiences.

Despite some difficulties, we found that the use of Change Laboratory method promoted an intense agency, for the ones who took part in it as well as to those who didn't take part in CL sessions. In addition, they mastered the method and began to use it in their professional practice.

This experience showed that, to ensure good conditions to apply, develop and consolidate CL methodology, a very structured negotiation phase is necessary with local as well as intermediate and higher management levels.

References

Bardin, L. (2009). Análise de Conteúdo. Lisboa: Edições 70.

Brasil. (2006). *Ministério da Saúde. Rede Nacional de Atenção Integral à Saúde do Trabalhador: manual de gestão e gerenciamento*. São Paulo: Ministério da Saúde.

Carvalho, G. (2013). A saúde pública no Brasil. *Estudos Avançados, 27*(78), 7–26. https://doi.org/10.1590/S0103-40142013000200002.

Dias, E. C., & Hoefel, M. G. (2005). O desafio de implementar as ações de saúde do trabalhador no SUS: a estratégia da RENAST. *Ciência & Saúde Coletiva, 10*(4), 817–827. https://doi.org/10.1590/S1413-81232005000400007.

Engeström, Y. (2015). *Learning by expanding: An activity-theoretical approach to developmental research*. New York: Cambridge Press.

Engeström, Y., Virkkunen, J., Helle, M., Pihlaja, J., & Poikela, R. (1996). The Change Laboratory as a tool for transforming work. *Lifelong Learning in Europe., 1*(2), 10–17.

Galdino, A., Santana, V. S., & Ferrite, S. (2012). Os Centros de Referência em Saúde do Trabalhador e a notificação de acidentes de trabalho no Brasil. *Cadernos de Saúde Pública, 28*(1), 145–159. https://doi.org/10.1590/S0102-311X2012000100015.

Sanchez, M. O., Reis Araújo, M., Cruz, A. L. S., & Ferreira, M. P. (2009). Atuação do CEREST nas ações de vigilância em saúde do trabalhador no setor canavieiro. *Saúde e Sociedade, 18*(1), 37–43. https://doi.org/10.1590/S0104-12902009000500006.

Vieira da Silva, L. M., & Almeida Filho, N. (2009). Equidade em saúde: uma análise crítica de conceitos. *Cadernos de Saúde Pública, 25*(2), 217–226. https://doi.org/10.1590/S0102-311X2009001400004.

Virkkunen, J., & Newnham, D. (2013). *The Change Laboratory: A tool for collaborative development of work and education*. Rotterdam: Sense Publishers.

Part III
Developments, Learning and Challenges for Change Laboratory Interventions for Workers Health

Chapter 14
Contributions from the Change Laboratory to the Analysis and Prevention of Accidents' Model

Sandra Lorena Beltran Hurtado, Rodolfo Andrade de Gouveia Vilela, Ildeberto Muniz de Almeida, José Marçal Jackson Filho, Marco Antonio Pereira Querol, Raoni Rocha Simões, Eugenio Paceli Haten Diniz, Manoela Gomes Reis Lopes, Alessandro José Nunes da Silva, Marcos Hister Pereira Gomes, and Susana Vicentina Costa

Abstract Analysis and Prevention of Accidents' Model (MAPA) was structured with the objective of replacing old practices that condemn work accidents on the victim. Applied to different activities, MAPA was able to explain accidents as a product of multiple interacting factors. However, a few changes have been implemented on underlying conditions. Overall, the agency was limited to the teams carrying out the analysis. In order to address this limitation, an interdisciplinary group participated in three workshops intended to reconstruct and improve the MAPA by using the principles of the Change Laboratory (CL). This chapter aims to present a

S. L. B. Hurtado (✉) · R. A. G. Vilela · S. V. Costa
Department of Environmental Health, School of Public Health, University of São Paulo, São Paulo, SP, Brazil
e-mail: sandrabeltran@usp.br

I. M. de Almeida
Department of Public Health, Botucatu School of Medicine, São Paulo State University "Júlio de Mesquita Filho", Botucatu, SP, Brazil

J. M. J. Filho
Occupational Safety and Medicine Foundation Jorge Duprat Figueiredo – Fundacentro, Curitiba, PR, Brazil

M. A. P. Querol
Department of Agronomic Engineering, Federal University of Sergipe, São Cristóvão, SE, Brazil

R. R. Simões
Institute of Pure and Applied Sciences, Federal University of Itajubá, Itajubá, MG, Brazil

E. P. H. Diniz
Occupational Safety and Medicine Foundation "Jorge Duprat Figueiredo" – Fundacentro, Belo Horizonte, MG, Brazil

R. A. G. Vilela et al. (eds.), *Collaborative Development for the Prevention of Occupational Accidents and Diseases*, https://doi.org/10.1007/978-3-030-24420-0_14

new model of activity system of analysis and prevention of work accidents. The text aims to answer the question: what are the new dimensions brought in from MAPA's development? We will present the stages and challenges of the new model.

14.1 MAPA and its Evolution

A thorough analysis of organizational causes of an accident disrupts the traditional approach, which blames the victim and is restricted to immediate causes. However, when the diagnosis is the result of an expert's analysis, the actions learned are limited to the team that investigated the event. Without the involvement and empowerment of local actors, significant gaps remain in the operational experience feedback (Revuelta 2004), and the much-desired organizational learning (Levitt and March 1988) is not achieved.

Hence, analysis does not always engage a social commitment capable of replacing the old approach of safety with an innovative conception that can be called psycho-organizational or systemic. Three poles support this new model: (1) conceptual contributions from external researchers to the company, (2) contributions of Occupational Health and Safety (OHS) professionals of the company and/or of governmental organisms, and (3) expertise contributions from operators and managers of the system in question.

MAPA is the product of a group of researchers and OHS professionals learning in the context of a public policy project that took place between 2007 and 2009 (Almeida and Vilela 2010; Almeida et al. 2014). The project aimed at investigating work accidents without using the safety approach that blames the victim, common in Brazil (Vilela et al. 2004). Initially, this group analyzed work accidents by checking technical or managerial non-compliances, using a normative framework, or else using causes tree analysis (Binder and Almeida 1997), in particular when analyzing severe and fatal cases.

MAPA's analysis team went on studying and took contact with the activity ergonomics (Wisner 1995) and other approaches such as cognitive psychology (Clot 1999) and organizational analysis (Llory and Montmayeul 2010; Dien et al. 2012), passing to include organizational aspects in their investigations.

MAPA integrates a set of concepts already used in accident analysis with a strong influence of the work ergonomic analysis (Guérin et al. 1997), change analysis (Leplat 1984), and barrier analysis (Hollnagel 2008). Its implementation encourages the analysis team not to be restricted to the accident's immediate causes that, as a rule, break down the event to "unsafe" behavior by the operator or technical failures—the famous binomial human error/technical failure.

M. G. R. Lopes
Department of Biological Sciences and Health, Federal University of Amapá, Macapá, AP, Brazil

A. J. N. da Silva · M. H. P. Gomes
Workers' Health Reference Center - CEREST, Piracicaba, SP, Brazil

The model is a tool for extended analysis of events, which aims to understand in depth its systemic origins—those related to coexisting tasks, redoing tasks ill projected, production processes, and its material conditions—besides human factors: linking them to the operative modes causes. This is a way of bringing into discussion organizational aspects that have to be improved, such as time pressure, production goals pressure, division of labor, hierarchy, staff size, budget cuts, communication difficulties, project decisions, logistics, maintenance, production, resource management, and safety management. The main stages of MAPA are general work analysis, barrier analysis, change analysis, and conceptual expansion (see Sec. 14.5). The model's last stage is structuring recommendations to avoid new occurrences.

One of the limitations identified with MAPA is that the companies investigated did not adopt many of the recommendations at an organizational level. This limitation led the group to questioning how to approach recommendations and to rethink strategies for involving inside actors of the organization in all analysis and transformation processes.

The cultural-historical activity theory (CHAT) has recently contributed to analyzing human activity-related accidents in nuclear power plants (Yoon et al. 2016) and to the analysis of risky environments (Norros 2004). Nevertheless, these studies are limited to diagnoses, without intervening in the situations and modifying their activities.

The Change Laboratory (CL) (Engeström et al. 1996; Virkkunen and Newnham 2013), drawing on CHAT's conceptual principles, helps to achieve a systemic view and also a formative intervention designed for qualitative transformation and expansive learning within work activities. It offers concepts and tools that contribute to MAPA to expand its object from analysis to prevention of accidents.

To address the current limitations of MAPA, we proposed to use the CL concepts and tools to expand two dimensions of the method: from the analysis performed by specialists alone to a collective analysis that involves local agents and from monitoring only Brazilian regulatory standards' (NR) compliance toward a more comprehensive understanding of latent causes of accidents.

14.2 The Workshops to Redesign the MAPA

The proposals made in this chapter are based on the discussion of several users of MAPA, during three workshops held in July 2016, December 2016, and May 2017, with the objective of analyzing and redesigning the model. On average 20 people, among which researchers, professors, OHS area postgraduate students, Workers' Health Reference Center (CEREST) staff, and a representative from a unit of the Labor Prosecution Office (MPT), participated in the discussion groups. Each one lasted 6 hours. A summary of the themes addressed during the first workshop is:

- Positive experiences/progress using the MAPA
- Negative experiences/limitations

- Possible dimensions for development
- New tools and stages of MAPA
- Challenges in its application

A draft paper and some figures to illustrate the new model were created during the second workshop. This material was shared with all the participants after the second and the third meeting, and any of them could comment on it. The result of this process is this chapter.

Aiming to advance toward the redesign of MAPA, we carry out a modeling exercise of the new activity system (AS) (Engeström 1987, 2015) that would represent the application of the tool, using the dimensions proposed in the previous item as inspiration to do so. Then, we discuss the possibilities of work accident analysis, the possible dimensions present in an expanded MAPA, the stages foreseen in the new model, and the challenges for its application.

14.3 Possible Dimensions for MAPA Expansion

The expansion on how MAPA understands accidents should occur in four categories: systemic, temporal, social, and transformative agency dimensions. These four dimensions connect dialectically, that is, they complement one another and do not exist separately. Therefore, no dimension is more important than the other. We will discuss each of them.

14.3.1 Systemic Dimension

This dimension refers on how to understand the AS under analysis, that is, the system's elements and their relationships (see Fig. 2.1). Therefore, expansion in this dimension involves a theory to explain how these elements interact so that the system produces its results.

Although traditional MAPA already involved several elements of the activity (e.g., individuals, tools), there was no solid unit of analysis that integrated these elements explaining how they interact. In this sense, the expansion of this dimension was possible by the introduction of the AS model as the basic unit of analysis. This concept allowed the object of the interventions to cease being an isolated element to connect dialectically; it also endorsed an expansion of analysis from individual actions to an activity, which interacts dialectically between micro- and macro-level.

In this perspective, accidents are understood theoretically as undesired results, expected or not, from human activity, consequences of actions or lack of actions, and of the malfunctioning of a system. When there are incompatibilities among the elements of an AS or between different activity systems, some actions may be conducted inappropriately or neglected due to structural incompatibilities within these elements. These incompatibilities can be understood as the result of historical changes. Therefore, understanding the emergence of contradictions that lead to accidents requires a historical analysis.

14.3.2 Temporal Dimension

This dimension refers to two aspects: (1) the chronological time spectrum of the actions under analysis and (2) the time necessary to conduct the learning actions during the interventions.

Expansion of the analyzed actions' time spectrum refers to the expansion of the object of analysis: from individual action to a group activity. It is essential to understand historically the radical changes that occurred in an AS's elements (Sewell and William 1996), the actions that led to the accident and their connection to the production process operation, the existence of contradictions between the elements (source of disturbances or unexpected events), and when and why these contradictions arose. A historical analysis of the emergence of the current AS could answer all these questions.

Secondly, the original MAPA approach involves analysis and recommendations. If recent assessments of MAPA results pinpoint learning actions as insufficient, it introduces the need for other stages, namely, negotiation, modeling, experimentation, implementation, and evaluation, which leads to increase the duration of interventions.

14.3.3 Social Dimension

The systemic and temporal dimensions described above require a social expansion of the interventions in the individual and in the community.

The issue of social expansion at the individual level relates to who conducts the learning actions. In a classic approach, specialists conduct these actions. However, it is a crucial trade-off between internal and external actors' views. An external view (specialist) brings in a different perspective for understanding the event but will have limited knowledge about how it really works. On the other hand, an internal point of view (local actor) can have a broader knowledge of the company, but a limited view on new perspectives (Cedergren 2013).

For local actors to adopt the suggested recommendations, they must first recognize the accident events as a real problem, then understand its causes and seek solutions. This learning process requires their involvement in all stages of the intervention.

Social expansion at the community level is directly related to the increase in the object's complexity. To be able to transform the newly expanded object, the team of investigation has to have new members when collecting data and transforming the new AS.

14.3.4 Transformative Agency Dimension

The success of interventions requires an expansion of a workers' power to act: they must be considered as actors of analysis and construction of solutions in prevention of accidents. There is a shift in their role, from insiders to agents of transformation. Such change requires their empowerment on how to conduct learning actions.

Learning creates space for confrontation and multi-voicedness (Engeström 2001), which is one of the most challenging modifications. It implies profound cultural changes, especially in authoritarian cultures—such as Brazilian workplaces. It also indicates change in how OHS professionals act, since historically they have used a behavioral approach, without understanding the actual frontline work. Otherwise, it means an expansion on how specialists are linked to the surveillance institutions. The leading change protagonists would become facilitators and promoters of healthy work environments.

This change has theoretical and practical consequences. It implies combining the role of inspectors with the role of negotiator and promoter of health and safety, according to the guidelines of the Occupational Health Surveillance (VISAT) (Ministério da Saúde 2009).

To detail a new system of accident analysis and prevention activity, we compared the activity systems of the original MAPA and MAPA Expanded, MAPA^{EX} (see Fig. 14.1). This representation is only a hypothesis that still needs to be improved, detailed, tested, and corrected. Its purpose is to serve as a starting point for the collective construction of MAPA^{EX}.

In short, participants of the workshops for redesigning MAPA concluded that the model should expand its unit of analysis in order to understand clearly the event, the work situation, and the AS where actions are inserted. Such a change in the tool creates the need to expand the learning actions involved in the model. In the original MAPA, the learning actions are related to the analysis made by the specialist. He/she collects documents, interviews people, confronts opinions and impressions about work, elaborates hypotheses, solves the puzzle, reaches a verdict, and returns a set of recommendations to the interlocutors. In that format, who learns the most is the analyst. The process allows neither a confrontation of actors' points of view nor the discussion about conflicts between production and safety. Even the most consistent recommendations are challenged by the legal department.

If consolidated, MAPA^{EX} could be characterized as a collaborative tool for organizational learning, in the collective operational experience feedback that can be triggered from several situations: fatal, severe, or major accidents; incidents with high potential threat to workers, the public or the environment; or crises of the

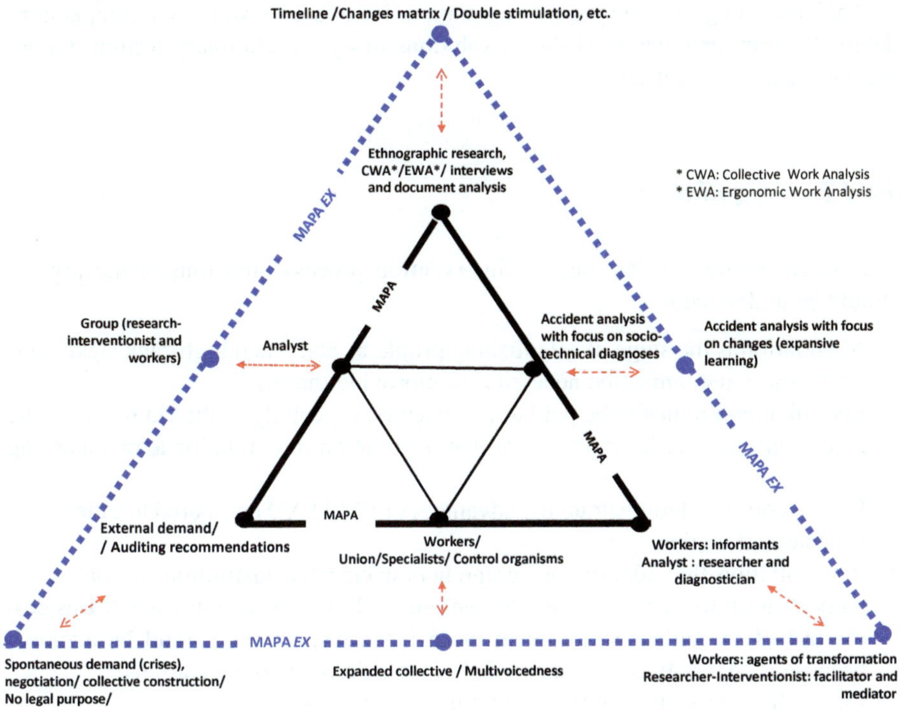

Fig. 14.1 Comparison of the MAPA and MAPA^{EX} activity systems. (Adapted from Engeström 2015)

traditional conflict between production and safety management. The basic principle of this new model is not to purposely establish fault of any nature, not even in higher hierarchical levels, but to promote the development and innovation within the activity. The expected learning is to achieve the understanding of an event as the result of local actors' constraints and how they conduct their actions according to crises, dilemmas, and conflicts. Hence the importance of carrying out the historical analysis of contradictions, searching for their origins and remodeling the system to create, develop, and test solutions that prevent not only new events but also provide a step toward safer and more efficient production.

14.4 Stages of MAPA^{EX}

One of the workshops' conclusions was that analysis and prevention needed to incorporate learning actions and include other previously unforeseen stages, such as negotiation, solution design, and experimentation. Even before negotiation, analysis itself arises from a demand that implicates a constraint for the organization, but where it is possible to initiate a social construction of changes.

The eight suggested stages are negotiation, introductory workshop, preparatory data collection, analysis workshops, solutions design, preliminary testing, implementation, and evaluation.

14.4.1 Negotiation

This stage is essential to start the intervention process, and thus, some aspects should be understood:

- Negotiation is the stage of establishing problems and when high-rank stakeholders ought to recognize the necessity of improving the AS.
- Special attention has to be paid to local actors, especially to the claimants of the investigation, in order to transform the negotiation into a dialogue and learning space.
- It is essential to demonstrate the advantages of MAPA^{EX} compared to other analysis methodologies.
- A negotiation that focuses on researchers asking the institution or company's authorization to carry out the intervention study is not recommended. Besides, the use of an authority argument as in labor inspections should be avoided, because it generally encourages a defensive attitude by organizations.
- Initial characterization of the event must be carried out.

14.4.2 Introductory Workshop

This workshop consists of an activity to facilitate the comprehension of workers about the method and conceptual principles of MAPA^{EX}. This stage will help participants from different hierarchical levels to come to a better understanding of what data are necessary for the analysis and their purpose. It is an exercise to discuss how the AS and other conceptual tools will be used to analyze the accidents. It lasts up to two hours and is recommended to hold a maximum of 20 participants.

14.4.3 Preparatory Data Collection

It is about collecting data that will serve to understand the usual work, the current and previous functioning of the AS, and the barriers and changes in the system that led to the event. Data collected will be organized and used as mirror data of the activity (see Sec. 2.8). OHS professionals in conjunction with other actors in the organization could perform this collection through the different techniques, detailed below.

Individual and Collective Interviews
These will seek information on historical changes in the management of production, equipment, and materials used over the years and the individual and social impact of the accident. The objective in this step is to collect data for analysis of the usual work, change analysis, and barrier analysis within the historical context of the company. Interviewees can be the person or people injured (where possible), other workers, managers of the involved areas, OHS professionals, trade unionists, or retired workers (dismissed or retired).

Collective work analysis (Ferreira 2015) may also be indicated: to gather a group of workers that perform the same task in which the event happened and obtain initial data to understand the usual work.

Direct Observation of Work Activities
Its objective is to understand the habitual activity in which the event happened, its variations, origins, innovations, and constraints in day to day work. It is still necessary to understand how the employees plan and perform the work, what they plan but cannot perform, how they act when faced with constraints, what they perform differently from the prescribed ways, and why and what they know and what they do not know when adopting a particular action.

Document Analysis
Through documents, it is possible to understand several aspects: (a) prescribed barriers (NR, manuals, preliminary risk analysis, work permits), (b) management decisions related to the event (contracts, budgets, schedules, service orders could help), and (c) OHS management (similar incident/accident records, accident investigation reports, communications of lessons learned, CIPA[1] minutes).

At the end of this stage, we suggest negotiating the composition of the team of actors who will plan and participate on the analysis workshops.

14.4.4 Analysis Workshops

The analysis workshops are planned, so that the participants can follow the cycle of expansive learning (see Fig. 2.4). As in CL, detailed planning of each meeting is essential to its success (Virkkunen and Newnham 2013) and therefore should include aspects as its objective, mirror data to be used as the first stimulus, and conceptual tools to be used as the second stimulus. Around eight two-hour workshops are necessary for event analysis, with one-week intervals and maximum participation of 20 people.

The learning action of questioning focuses on helping actors involved to recognize the existence of a problem. During this phase, participants should be provoked to define the underlying activity that led to the accident. In this analysis stage, we

[1] CIPA is an Internal Commission for the Prevention of Accidents, composed by representatives from the company and the workers and mandatory in every Brazilian company.

propose to use operational experience feedback from accidents and critical incidents in order to boost the learning of the group involved. Reconstruction and simulation of the accident situation can also be used, with the participation of team members from the sectors involved.

The action analysis focuses on knowing the problem. Such analysis may be necessary for a better understanding of actions and operative modes adopted to perform the tasks at the time of the accident. It means finding a way of representing what happened in the course of the action and/or operation where the subject, individual or collective, was in. Understanding the event at a level of action implies understanding its specific dynamics, that is, the proximal dynamics that gives the clues to a deepening of the distal aspects (Hale et al. 2007).

The analysis sessions should cover the usual work, the change analysis, the barrier analysis, and conceptual expansion and should end with a brief description of what happened (what, with whom, where, and when) and the historical contradictions that contributed to the accident.

Analysis of the Usual Work

It is based on concepts of the activity ergonomics to understand the context of the tasks (without the accident) at the level of the situated action (Suchman 1987). It includes the steps, available resources, variations, and many more constraints of actual work and how individuals deal with these constraints, strategies, and operative modes mobilized.

Identifying and modeling the activity in which the accident occurred within the triangle of the AS can be an essential second stimulus, for a systemic understanding of the contradictions within each component or between subjects, tools, rules, community, and division of labor and object.

Analysis of Changes

This aims to identify the differences between the work with and without accident and simultaneously explore the systemic conditions associated with the origins of these changes. This analysis makes it possible to compare what happened at the level of action, what triggered the accident situation, what changes occurred in the elements of the system, and what motivated the emergence of these events.

Understanding an accident cannot be restricted to the identification of contradictions in a present situation, because it is essential to understand the historical development of the AS to think about solutions. What historical events provided the contradictions identified in the present situation? What structural tensions exist today and what are their origins?

Researchers can use three concepts as a second stimulus to assist participants in achieving these learning objectives:

- The timeline of critical historical events that changed the relationship of AS elements (Sewell and William 1996) grouping such events in time periods.
- The bow tie model (Hale et al. 2007) to help the participants identify proximal causes and latent conditions.

- The changes matrix (Virkkunen and Newnham 2013), which seeks to analyze changes in the object, subject, instruments, community, division of labor and rules (horizontal axis) throughout time periods identified previously in timeline (vertical axis).

Barrier Analysis
This comprises the analysis of potential dangers and of which barriers—prevention, monitoring, and/or protection—are present or absent in the system. It is critical to understand if these barriers existed, worked, or failed. What explains the inexistence or failure of these barriers? The Swiss cheese model (Reason 2000) and the barrier matrix (Almeida and Vilela 2010) can serve as a second stimulus in understanding failures or lack of prevention and protective barriers.

Conceptual Expansion
This consists of reanalyzing the information collected with the support of concepts from different areas of knowledge already used in systemic accident analysis. Some questions stimulate the explanation of operators' behaviors, which are usually pointed out in companies' investigations, as the ultimate explanation of what happened. Emphasis should be on understanding the behaviors of the injured, rather than judging them, questioning what were the reasons, during the accident situation, for the injured to consider acceptable acting as they did.

From the cognitive point of view, were there any traps or situations of surprise that made it difficult for the operator to understand and control the variations? What concepts help to explain the situated behavior of the operator? Concepts of cognitive psychology such as the cognitive trap or predictable bypass can be offered as a second stimulus, to explain aspects not yet clarified of the reasons that led to a particular behavior of the worker.

14.4.5 Modeling the Solutions

Based on the analysis phase, at this stage participants think about possible changes in the elements of the AS and begin to design new solutions for working without accidents. Positive and successful experiences, as well as the AS model, can be used again as first stimulus to identify which elements should change and what implications these changes would cause on the system. Modeling should predict interferences in other tasks and other programs and sectors of the company. Appropriation of innovation must be systemic, that is, all elements of the system must be prepared for this innovation. The user of such innovation must participate in the whole project. Intermediate objects can be used as second stimuli during this phase. They can be material (figures, prototypes, simulators) or immaterial (new rules, spreadsheets, schemes).

Between two and three sessions are necessary for the design of solutions. The stages of analysis and design of the solutions should be held in weekly sessions, followed by an interval of about three months, for practical testing. After the interruption, the follow-up phase can be carried out in three to four sessions for testing, implementing, and evaluating the solutions implanted.

14.4.6 Testing the New Model

At this stage, the focus is to bring the abstract ideas raised in the modeling step to a more concrete level and closer to becoming real. In the sessions where intermediate objects are used, it is vital to consider workers' know-how and have space for expression. Participants must be the protagonists, and the interactive dialogue for redesign must guarantee a multi-voiced perspective.

14.4.7 Implementing the New Model

During this stage, sessions will have longer intervals. Their objective is to assess: Were the solutions implemented? What were the difficulties in implementing those solutions? How were the difficulties overcome? Which have not yet been solved? With this change, have new problems appeared? Which efficacy indicators are being used to evaluate the new solution? Are the indicators enough? Would others be needed?

The AS is used again to analyze whether the contradictions that existed remain or not, and to find the new contradictions or those that were not perceived or were discrete, becoming relevant after the modeling. It may be necessary to go back to the examination phase with new interactive design dialogues, new intermediate objects, and broadening the network of actors involved (Virkkunen and Newnham 2013).

14.4.8 Evaluating and Consolidating the New Model

The new model should be evaluated according to the criteria and indicators raised in the previous stage. When the solution is valid, it is assessed regarding its extension and possibility of use in other sectors, other facilities, and other organizations from the same sector or with similar activities. Consolidation phase is replacing previous practices and the solidity of the new practice.

The eight stages proposed, with their main characteristics, are summarized in Table 14.1

Table 14.1 Summary of expanded MAPA stages

Stages	Objectives	Strategy/methods	Participants
1. Negotiation	To arrange the scope, objectives, and principles of research. To outline the collaborative process. To foster trust and cooperation. To establish commitment and ethical rules. To discard legal bias. To define the term of cooperation. To create the supporting team. To establish access conditions for exploiting managerial decision-making related to the event	1E[a]: What is the point of view of the organization and actors about the event? 2E[b]: Study proposal. Benefits for the company. In what accidents could be applied? Was it an active demand from the company or it had to be stimulated?	Researchers-interventionists Managers OSH professionals *Optional*: Prosecutors Labor inspectors Union representatives
2. Introductory workshop	Disclosing the main concepts of the methodology to the participants	Practical exercises with examples of activity systems	Workers from different hierarchical levels
3. Preparatory data collection	Preliminary information about the event, the activity, and the organizational context Preparation of data that will serve as a mirror for the workshops To meet actors and build trust for cooperation To create hypotheses for explanation of systemic contradictions	Individual and collective interviews Collective analysis of work Direct observation of activities with video and photo registration Documents analysis	Researchers-interventionists and internal actors. Internal actors should ideally collaborate in data collecting. Data must be validated with the participants
4. Analysis workshops	To understand the usual work; the contradictions between the elements of the AS that influence the workers' behavior; the dynamics of the event in the context of the situated action To identify the differences between working with and without the accident, potential hazards and barriers for prevention, monitoring and protection present or absent in the system To explore systemic conditions associated with organizational changes	1E: Presentation of collected data about the activity or about the event to start the discussion, followed by the question: is there a problem? 2E: AS model; the changes matrix; bow tie model; Swiss cheese model or barrier matrix; timeline; concepts (cognitive trap, foreseeable deviation)	Researchers-interventionists and volunteer participants in workshops

(continued)

Table 14.1 (continued)

Stages	Objectives	Strategy/methods	Participants
5. Workshop: modeling the solutions	To elaborate solutions at a systemic level that overcome the contradictions identified. To sketch an action plan	1E: What changes in the elements of the AS are necessary? Next steps to develop and test solutions 2E: AS model, prototypes, pilot tools, simulators, spreadsheets, schemes	Researchers-interventionists and volunteer participants in workshops
6. Testing the new model	To bring abstract ideas raised in the modeling step to a more concrete level	Solutions should be negotiated at decision-making levels Changes modeled in the real activity could be tested in a pilot unit	Volunteer participants in workshops Other internal actors
7. Workshop: evaluating the new model	To evaluate and adjust developed solutions	1E: Have the solutions come to the expected results? What difficulties have to be overcome? Are other measures necessary? 2E: The same models used in the workshop of modeling solutions	Researchers-interventionists and volunteer participants in workshops
8. Consolidating the new model	To analyze if the solution is sufficient enough to be extended to other sectors, departments, or facilities	Replacing previous practices and the consolidation of the new practice. The new model now tested and approved will take effect and serve as an example for others?	Researchers-interventionists and negotiation group

[a]1E: First stimulus [b]2E: Second stimulus

14.5 The Challenges of MAPA^{EX} Development

Despite the significant advances achieved during the workshops, there are still considerable challenges for MAPA^{EX} design and implementation. At least two significant challenges are identified for the model.

14.5.1 From Abstract to Concrete

The first challenge of MAPA^{EX} is its implementation. The ongoing design of the model needs to take into consideration its practical application, a translation of what the new principles mean regarding methods, procedures, and tools. According to Lindberg et al. (2010), studies that increase our understanding of the efficiency of different investigation methods and the actual effects of these investigations on preventive actions are necessary. Cedergren (2013) raises a similar point regarding the explanations of successful implementation as a possibility for future research.

A contribution to this approach is the systematization, in a structure similar to CL's, of a collective process of analysis of the accidental events, understood as being fruit of contradictions present in the AS. This structure would allow the expansive learning (see Fig. 2.5); nonetheless, other phases at the beginning of the cycle will be necessary. The figure below attempts to detail the steps for analysis and collective construction of solutions after the occurrence of an event (see Fig. 14.2).

The new MAPA adds elements in its AS. This change requires the expansion of object (effective implementation of organizational changes), the inclusion of participants (involving local actors in the process of analysis), and instruments (new tools for collecting and analyzing data). Besides, it would take a change in rules (new spontaneous demands without legal purpose), an increased community (multi-voiced perspective), and changes in division of labor (researchers-interventionists as mediators and workers as agents of transformation).

14.5.2 Practicality vs. Qualification

Another challenge of the MAPA^{EX} is the requirement of an even better understanding of concepts and more in-depth knowledge of the activity under intervention. Tool applicators should have the expertise and have participated in a course on the methodology of accident analysis with emphasis on activity ergonomics, expansive learning theory, CHAT, and CL. Besides, the application of MAPA^{EX} increases the time necessary for more in-depth and collective analysis of the actors involved in the AS.

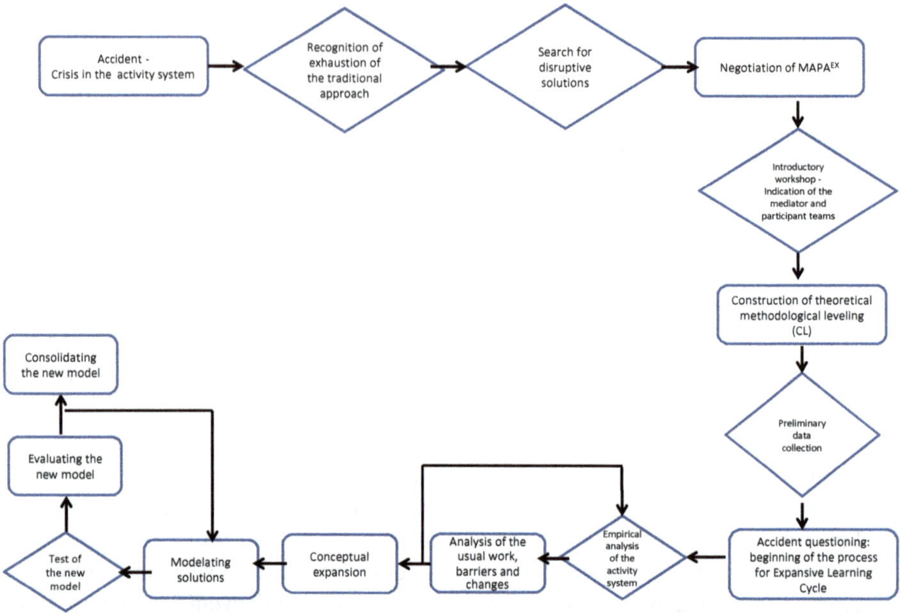

Fig. 14.2 Expanded MAPA flow chart

Original MAPA use already meant an increase in analysis time, which goes against increasing demands from society for more straightforward analysis. MAPA^{EX} can further aggravate the contradiction between production goals of accident analysis and higher-quality interventions with intended far-reaching results. One of the strategies is to develop tools that meet different demands or different levels of performance. For example, simpler cases could use basic instruments, while more complex cases would be selected as sentinel events and analyzed through MAPA^{EX}.

We envision that MAPA could be used, potentially, in inspection and prevention activities or as an analysis tool for OHS professionals. However, within companies there are strong legal pressures for simpler analyses that preferably identify workers as responsible for their accidents. A company's agreement to perform an organizational and systemic study would require that the tool be applied exclusively for preventive purposes, as it is with current plane crashes' investigations, and that it will not be used solely or mainly to identify guilt.

On the other hand, public institutions for surveillance and regulation of labor can only act within the strict limits of what is legally indicated. In those cases of noncompliance to legal norms, the auditor has to issue a fine to the company, which creates an impasse situation. If the company collaborates, it can learn with help from MAPA^{EX}, but it could lead to it being penalized for the collaboration. If the company does not collaborate, it does not learn, and safety regulators continue to suffer from insufficient analyses.

MAPA^{EX} is still in the developmental phase of testing the last stages suggested in this chapter. When implemented, it is very likely that new contradictions will emerge in the activity of work accident prevention. It is only by putting the new tool to test that the challenges will emerge and be solved.

This study was financed in part by the Coordenação de Aperfeiçoamento de Pessoal de Nível Superior—Brasil (CAPES)—Finance Code 001.

References

Almeida, I. M., & Vilela, R. A. G. (2010). *Modelo de Análise e Prevenção de Acidentes de Trabalho – MAPA*. Piracicaba: CEREST.

Almeida, I. M., Vilela, R. A. G., Silva, A. J. N., & Beltran, S. L. (2014). Modelo de Análise e Prevenção de Acidentes – MAPA: ferramenta para a vigilância em Saúde do trabalhador. *Ciênc Saúde Colet, 19*(12), 4679–4688. https://doi.org/10.1590/1413-812320141912.12982014.

Binder, M. C. P., & Almeida, I. M. (1997). Estudo de caso de dois acidentes do trabalho investigados com o método de árvore de causas. *Cad Saúde Pública, 13*(4), 749–760.

Cedergren, A. (2013). Implementing recommendations from accident investigations: A case study of inter-organisational challenges. *Accident Analysis and Prevention, 53*, 133–141. https://doi.org/10.1016/j.aap.2013.01.010.

Clot, Y. (1999). *La fonction psychologique du travail*. Paris: PUF.

Dien, Y., Dechy, N., & Guillaume, E. (2012). Accident investigation: From searching direct causes to finding in-depth causes - problem of analysis or/and of analyst? *Safety Science, 50*(6), 1398–1407. https://doi.org/10.1016/j.ssci.2011.12.010.

Engeström, Y. (2001). Expansive learning at work: Toward an activity theoretical reconceptualization. *Journal of Education and Work, 14*(1), 133–156. https://doi.org/10.1080/13639080020028747.

Engeström, Y. (2015). *Learning by expanding: An activity-theoretical approach to developmental research*. New York: Cambridge Press.

Engeström, Y., Virkkunen, J., Helle, M., Pihlaja, J., & Poikela, R. (1996). The Change Laboratory as a tool for transforming work. *Lifelong Learning in Europe, 1*(2), 10–17.

Engeström, Y. (1987). *Learning by expanding: An activity-theoretical approach to developmental research*. Helsinki: Orienta-Konsultit.

Ferreira, L. L. (2015). Análise coletiva do trabalho: quer ver? Escuta. *Revista Ciências do Trabalho, 4*, 125–137.

Guérin, F., Laville, A., Daniellou, F., Duraffourg, J., & Kerguelen, A. (1997). *Comprendre le travail pour le transformer – La pratique de l'ergonomie*. Lyon: ANACT.

Hale, A. R., Ale, B. J. M., Goossens, L. H. J., Heijer, T., Bellamy, L. J., Mud, M. L., et al. (2007). Modeling accidents for prioritizing prevention. *Reliability Engineering and System Safety, 92*(12), 1701–1715. https://doi.org/10.1016/j.ress.2006.09.025.

Hollnagel, E. (2008). Risk + barriers = safety? *Safety Science, 46*(2), 221–229. https://doi.org/10.1016/j.ssci.2007.06.028.

Leplat, J. R. (1984). Analysis of human errors in industrial incidents and accidents for improvement of work safety. *Accident Analysis and Prevention, 16*(2), 77–88.

Levitt, B., & March, J. G. (1988). Organizational learning. *Annual Review of Sociology, 14*, 319–338. https://doi.org/10.1146/annurev.so.14.080188.001535.

Lindberg, A. K., Hansson, S. O., & Rollenhagen, C. (2010). Learning from accidents – What more do we need to know? *Safety Science, 48*(6), 714–721. https://doi.org/10.1016/j.ssci.2010.02.004.

Llory, M., & Montmayeul, R. (2010). *L'accident et l'organisation*. Bordeaux: Préventique.

Ministério da Saúde (BR). (2009). Portaria n° 3.252, de 22 dezembro 2009. Aprova as diretrizes para execução e financiamento das ações de Vigilância em Saúde pela União, Estados, Distrito Federal e Municípios e dá outras providências. *Diário Oficial da União*. 22 dez 2009. Seção 1, pp. 45–50.

Norros, L. (2004). *Acting under uncertainty. The core-task analysis in ecological study of work.* Espoo: VTT publications.

Reason, J. (2000). Human error: Models and management. *BMJ, 320*(7237), 768–770. https://doi.org/10.1136/bmj.320.7237.768.

Revuelta, R. (2004). Operational experience feedback in the World Association of Nuclear Operators (WANO). *Journal of Hazardous Materials, 111*(1–3), 67–71. https://doi.org/10.1016/j.jhazmat.2004.02.025.

Sewell, J. R., & William, H. (1996). Historical events as transformation of structures: Inventing revolution at the bastille. *Theory and Society, 25*, 841–881. https://doi.org/10.1007/BF00159818.

Suchman, L. (1987). *Plans and situated actions: The problem of human/machine communication.* Cambridge: Cambridge University Press.

Vilela, R. A. G., Iguti, A. M., & Almeida, I. M. (2004). Culpa da vítima: Um modelo para perpetuar a impunidade nos acidentes do trabalho. *Cadernos de Saúde Pública. 20*(2), 570–579. https://doi.org/10.1590/S0102-311X2004000200026.

Virkkunen, J., & Newnham, D. S. (2013). *The Change Laboratory: A tool for collaborative development of work and education.* Rotterdam: Sense Publishers.

Wisner, A. (1995). *Réflexions sur l'ergonomie (1962–1995).* Toulouse: Octarès.

Yoon, Y. S., Ham, D. H., & Yoon, W. C. (2016). Application of activity theory to analysis of human-related accidents: Method and case studies. *Reliability Engineering and System Safety, 150*, 22–34. https://doi.org/10.1016/j.ress.2016.01.013.

Chapter 15
Learning in and from Change Laboratory Interventions for Developing Workers' Health in Brazil

Rodolfo Andrade de Gouveia Vilela, Marco Antonio Pereira Querol, Amanda Aparecida Silva-Macaia, and Sandra Lorena Beltran Hurtado

Abstract This chapter summarizes the main results of the ten formative interventions presented in this book. We discuss the strategies used by our research group to support the learning process of the key theoretical and methodological tools used in the Change Laboratory method. We continue describing the role of demand, negotiation, the process of data collection, and the setting of the sessions. We summarize the hypothesis of contradictions and main results achieved so far in the ten interventions. We also reflect about the main learning challenges, the key innovations, and what were the key learning aspects in our research group. We argue that the CL method allowed not only the expansion of the work activities under intervention presented in this book but also the expansion of our research-interventionist activity, contributing to a broader understanding of who and how should be involved in interventions for developing workers' health and safety. The method contributed toward a more systemic and historical approach in which interventions should be collaboratively constructed rather than imposed by an expert. Moreover, thanks to the theoretical tools used in the method, the findings could be generalized to other cases, allowing the new locally produced concepts to be applied in different settings.

R. A. G. Vilela (✉) · A. A. Silva-Macaia · S. L. B. Hurtado
Department of Environmental Health, School of Public Health,
University of São Paulo, São Paulo, SP, Brazil
e-mail: ravilela@usp.br

M. A. P. Querol
Department of Agronomic Engineering, Federal University of Sergipe,
São Cristóvão, SE, Brazil

R. A. G. Vilela et al. (eds.), *Collaborative Development for the Prevention of Occupational Accidents and Diseases*, https://doi.org/10.1007/978-3-030-24420-0_15

15.1 Introduction

The motivation of our group to learn and use the Change Laboratory (CL) method (Engeström et al. 1996) was related to the need to address the shortcomings of the intervention methodologies already used to deal with health and safety in the workplace (see Chap. 1). These limitations determined the development of three strategic components:

- Expansion of practitioners' agency over the process by going from the specialist perspective to one of collective work
- Expansion of the understanding of problems' causality, by going from a linear, deterministic view (uni- or pauci-causal type, generally searching for a culprit), or even a factorial view, to a dialectical, historical, and systemic causality
- Expansion of the learning process, by going from an interventionist-specialist approach that analyzes, learns, and returns a diagnosis/recommendation to an integrated, expansive, and collective learning approach

All these efforts toward expansion, however, would be in vain if local actors facing problem situations do not recognize that these problems need to be solved. This means that transformation cannot be a desire of a researcher, of a public agency for accident surveillance and prevention, but must be constructed as a shared object between the activity system developed by interventionists and the activity system of the productive organization (see Fig. 15.1).

As seen in Chap. 2, CL involves a set of principles, models, theories, concepts, and methods that aim to facilitate expansive learning (Engeström 1987, 2015). It is a learning platform that combines spaces of emotional involvement and concrete

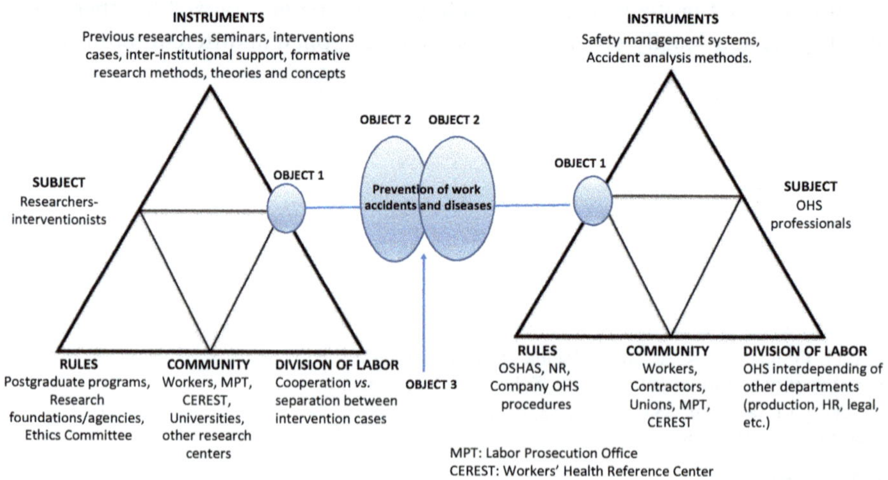

Fig. 15.1 Shared object of the interventionist-researchers' activity and the institution where the intervention happened. (Adapted from Engeström 2009)

learning tools that help to understand and resolve historical contradictions within and among the elements of the activity system under intervention. The method puts interventionists and participants in the same "boat." Our experience, portrayed in this book, shows that this instrument, while possessing a great set of virtues, is not a panacea.

An interventionist methodology needs to adapt to the specific cultural context of each country. Aspects such as hierarchical relationships and the social acknowledgment of the problem, in a more local context, determine the engagement of different actors in work transformation. This means that the response will always be unique and dependent of the local reality, the interest of the demanding organization, and the team's abilities in conducting the intervention process in this specific context.

When interventionists-researchers first came across the CL method, they encountered a number of challenges, among them the need to train a team capable of conducting interventions with the method and the lack of recognition of this methodology by the community. This has led to the question of how to deal with the need of intervention in organizations, how to reach hierarchical upper levels of organizations and which negotiation strategies are most suitable to each situation.

The main challenges that our research team had to face in order to develop the interventions with CL were:

- Mastering the cultural-historical activity theoretical approach, which has some differences and similarities with theoretical basis of activity ergonomics, already known by some members of the group.
- Mastering expansive learning theory, which offers to researchers a set of pedagogical tools that favor participants' learning.
- Developing skills to articulate a broadened research community into a network of social and political actors.
- Developing skills to negotiate demands from organizations interested in transforming their activity systems, including negotiation skills with managers, to provide a favorable environment for formative and collaborative intervention research.
- Developing national and international network to create a space for learning, dissemination of scientific results, and development of the methodology itself.

To achieve these challenges, first we had to acknowledge that expansive learning requires the expansion not only of the object but also other elements of the activity system. This process requires a change in the role of the researcher-interventionist, in the theoretical-practical instruments, in the role of the community (participants are no longer research informants but primordial subjects of analysis and change) and in research rules, which include greater attention and care in negotiations. It also requires training researchers and fine-track planning and sharing with the organization's managers, etc. The actions in this context were not trivial challenges for a team that seeks to aggregate in its repertoire concepts and tools of activity ergonomics (Guérin et al. 1997) and Paulo Freire's liberating education (1993) together with the cultural-historical activity theory (Engeström 1987, 2015).

Initially the goal was to build local skills to conduct interventions. Other actions sought to disseminate the methodology to the community. Another set of actions dealt with the conduction and continuous evaluation of the interventions with the organizations. Implementing the intervention on the other hand required cooperative forms of division of labor, continued supervision, and international exchange.

As we have tried to highlight in previous chapters, our experience has faced these challenges in different contexts.

As it will be described below, these actions are not static; on the contrary, they are interconnected in a back and forth movement placing the activity system around the object of the intervention.

15.2 Qualifying in CL Method

The starting point of this journey was in 2011, the contact of our research group with Dr. Marco A. P. Querol, who did his PhD at the University of Helsinki, Finland, using an activity theoretical approach. Through this, we were able to start a promising cooperation with researchers from the Center for Research on Activity, Development and Learning (CRADLE) of the University of Helsinki.

In 2012 we held the first training course (40 hours) in the CL method at the School of Public Health - University of São Paulo (FSP/USP). It was taught by Professor Jaakko Virkkunen, who presented theoretical-conceptual aspects and examples of intervention. It was an expositive course, and the group experienced learning difficulties due to lack of practical experience with the methodology.

At the time of this course, professor Virkkunen was finishing the book *The Change Laboratory: A Tool for Collaborative Development of Work and Education* (Virkkunen and Newnham 2003). We became interested in publishing a Brazilian edition, which came to light 1 year later. This reference material in Portuguese was vital to the process of learning and applying the method. The course and the book inspired the team to plan and conduct the first three interventions using CL: Workers' Health Reference Center (CEREST) Piracicaba (see Chap. 13), airport construction (see Chap. 9), and the FSP health center (see Chap. 10). After this first course, three team members took part in a summer course and internships at the University of Helsinki between 2012 and 2013.

We conducted a short course at CEREST Piracicaba, coordinated by Professor Marco Querol, with the main goal of presenting the methodology to the institution's employees. At the same time, we started data collection and planning of CL conduction in the health center. These two cases and the ones that followed were subject to continued supervision of Professor Marco Querol regarding planning, negotiation, evaluation, and even conduction of the sessions.

This follow-up of the interventions was also performed by Finnish Institute of Occupational Health (FIOH) Doctor Laura E. Seppänen during the 5 months between 2015 and 2016 in which she stayed as visiting researcher in Brazil, for the research project—"Change Laboratory for developing health at work." Under her coordination we undertook the "Seminars of Intervention Studies/CL." She participated in the first four meetings before returning to Finland, and the seminars are still happening, organized by Amanda Silva Macaia, postdoctoral student, in collaboration with Professor Rodolfo Vilela. The seminars seek, through a forum between researchers and students, to enhance theoretical and practical understanding of CL, to share information, to provide mutual support, and to evaluate and learn from the interventions carried out. Each meeting includes the discussion of a previously read article or theoretical chapter and a presentation and discussion of at least one case of intervention with CL. Eight to fifteen people attend the seminars, including intervention subgroups' representatives and people interested in the methodology. Table 15.1 presents information on the contents and cases covered.

The seminars provide broadened teams for discussing each intervention subgroup, reinforcing the idea advocated by the methodology that projects should not be conducted by just one person.

The seminars have become important, strategic spaces for the mastery of theoretical aspects of the methodology, as well as exchanging experiences on planning and implementing interventions. In addition to the Seminars for Intervention Studies/CL, the group organizes an annual general seminar to discuss the experiences, to which we invite external speakers.

We have described the actions taken for the development of local competence, conduction, and continual evaluation of the interventions. In order to incorporate and reach these goals, courses and disciplines on the methodology were offered, discussing the methodology applied with focus on work-related accident and injuries prevention in the Brazilian context.

In 2015 a discipline was created within the postgraduate program of the FSP/USP, "Change Laboratory Method for Research and Development of Teams and Professionals." This was done in collaboration with Professor Jaakko Virkkunen, with the participation of the visiting researcher Laura Elina Seppänen and support of Professor Rodolfo Vilela, postdoctoral and PhD students. The discipline structure combined theoretical concepts with discussion of intervention cases in progress at that time and required the completion of pre-course tasks.

The new pedagogical strategy adopted in the discipline showed its adherence to the precepts of expansive learning (Engeström 2015). From the requirements and hardships faced in the conduction of CL (need state) and the offer of tasks to the participants (first stimulus), the theoretical concepts (second stimulus), previously perceived as abstract, gradually began to be internalized and mastered by the participants.

Table 15.1 Theoretical content and cases discussed in the meetings of the intervention study seminars

No.	Date/place	Theoretical content (theme and readings)	Practical content
1	Feb 24, 2016 FSP/USP	Double stimulation. Putting Vygotsky to work: the Change Laboratory as an application of double stimulation (Engeström 2007)	Discussion of the term "intervention" in Brazilian context Planning of four interventions: Limeira Municipal Committee for Eradication of Child Labor and Adolescent Work Protection/ COMETIL; Adolescent Socio-educational Assistance Center/CASA Foundation; school; and School–Health Center/CSE
2	Mar 23, 2016 Piracicaba - CEREST	Theoretical concepts. Putting Vygotsky to work: the Change Laboratory as an application of double stimulation (Engeström 2007)	Cases discussion: Hemodialysis on Bauru Hospital; prevention of risks in construction called "work in heights"; and urban cleaning Question for debate: How do researchers and companies dialogue when negotiating and preparing for CL?
3	Apr 27, 2016 FSP/USP	The construction of agency. Examining developmental dialogue: the emergence of transformative agency (Heikkilä and Seppänen 2014); Generating transformative agency among horticultural producers: an activity theoretical approach to transforming integrated pest management (Vänninen et al. 2015); O papel da dupla estimulação e da agência para entender as contradições em obras de construção de um aeroporto (Lopes et al. 2016).	Ongoing interventions using the agency concepts
4	May 11, 2016 Piracicaba - CEREST	Double stimulation, with focus on expansive learning	Examples of double stimulation uses (mirror data used, first and second stimuli) on the CL negotiation and session phases
5	Jun 22, 2016 FSP/USP	The role of management in interventions. Shared challenge for learning: dialogue between management and front-line workers in knowledge management (Ahonen and Virkkunen 2003); design as a mutual learning process between users and designers (Béguin 2003)	How to approach the view of managers and other actors? How to build a common object? What is the context needed for managers' involvement in the interventions?
6	Sep 2, 2016 FSP/USP	Hypotheses of contradictions and development of perspectives for projects in progress; empirical data collection plan	Cases planning for the CL discipline offered in October 2016
7	Apr 6, 2017 FSP/USP	Contradiction. Studies of expansive learning: foundations, findings, and future challenges (Engeström and Sannino 2010)	Cases discussion: Electric sector; CEREST; and railway company. How to plan the learning process? Was there growth in actors' agency? Which primary and secondary contradictions were identified by actors?
8	May 19, 2017 FSP/USP	Contradiction. Studies of expansive learning: foundations, findings, and future challenges (Engeström and Sannino 2010)	Cases discussion: Bauru Hospital; and health network in Piracicaba. Theoretical questions guided the cases discussion. A collective planning of a meeting with managers from the Department of Health regarding the context of the Health Network CL also took place

No.	Date/place	Theoretical content (theme and readings)	Practical content
9	Jun 19, 2017 FSP/USP	Contradictions of capitalism. Contradictions of high-technology capitalism and the emergence of new forms of work. In: Learning and expanding with activity theory (Mietinen 2009)	Cases discussion: CASA foundation; urban cleaning; and COMETIL Cases were discussed regarding the theme "identification of the primary and secondary contradictions of the activity systems."
10	Sep 21, 2017 FSP/USP	Information drafting on the cases for the upcoming CL discipline. The discipline was structure to plan the actions in three chosen interventions	Cases discussion: Clothing making; electrical sector; railway company; CASA Foundation; Bauru Hospital; Family Health Program/PSF; urban cleaning; CEREST; and health network. Demand, unit of analysis, object, main activity system problems, hypothesis of historical contradictions, current phase of the cycle and continuity plan, main challenges, and struggles faced by researchers
11	Nov 28, 2017 FSP/USP	Agency. The emergence of learners' transformative agency in a Change Laboratory intervention (Haapasaari et al. 2016)	Cases discussion: COMETIL; professional rehabilitation
12	Feb 6, 2018 FSP/USP	The activity system mediators: production, consumption, distribution, and exchange. The evolution of activity (Engeström 1987, 2015)	Cases discussion: Urban cleaning; PSF. Discussion on strategies used, planning next stages, debate/analysis regarding theoretical concepts
13	Apr 4, 2018 FSP/USP	The activity system mediators: Production, consumption, distribution, and exchange (Marx 2011)	Cases discussion: Urban cleaning; railway company (double stimulation)
14	Jun 7, 2018 FSP/USP	Contradictions of contemporary capitalism. Work organization and occupational health in the contemporary capitalism (Dias and Lima 2014)	Cases discussion: CASA Foundation; endemic diseases control agents
15	Sep 19, 2018 FSP/USP	1- Rapprochement between Vygotsky and Paulo Freire (Alves 2007, 2012) 2- Agency	Cases discussion: Professional rehabilitation; health care network
16	Nov 30, 2018 FSP/USP	History after Vygotsky (Scribner 1997).	Cases discussion: PSF; endemic diseases control agents (negotiation and data collection stages)
17	Mar 13, 2019 FSP/USP	Relational agency. Revealing relational work (Edwards 2007)	Cases discussion: Urban cleaning; Osasco school; and workers' health interventions
18	May 22, 2019 FSP/USP	Transformative agency in formative interventions (Sannino et al., 2016)	Cases discussion: Hospital of Federal University of São Carlos (UFSCar); and Women's Health Reference Center (MATER) at Ribeirão Preto
19	Jul 2, 2019 FSP/USP	Paulo Freire's method (Feitosa, 1999)	Cases discussion: Endemic control diseases agents; column surgery for construction workers

These initial years of experience motivated us to offer the first Brazilian version of the discipline in October 2016, adapted to the country's context and to the group's needs. We have increased the course workload and formed a group of instructors.

The process of planning the discipline occurred in a collaborative way between the group of professors/instructors and was open to the students' learning needs. Because of this, we decided for continual evaluation of the course, and several changes were made during its development. We used CL toolkit strategies to shape the course, which was very well evaluated by the 23 participating students.

The disciplines proved to be an essential opportunity to improve the group of teachers' knowledge on the methodology, insofar as even while they taught, they learned. It was noteworthy that recurring participation in the course as a tutor or as a participant of the discipline, associated with practice in the cases, improved the team's understanding on CL's theoretical and methodological framework.

In 2017 the discipline was offered again with the Finnish professors, plus the participation of Professor Emeritus Reijo Miettinen, who is also a member of CRADLE. He ministered a 16-hour seminar prior to the course, approaching the conceptual part by discussing eight key concepts. The planning and conduction of the interventions was coordinated by Professor Jaakko and Professor Laura in 40 hours, and case discussions in subgroups were enriched by the presence of people involved in ongoing CLs.

In 2018, the Brazilian team took over the course, preserving the same structure as the preceding year. The conceptual part was given by Professor Marco Querol and the case planning by Professor Rodolfo and his team.

It seems worthy to highlight that the group was already experienced since some of its members were trained in activity ergonomics (Guérin et al. 1997), a method that uses problem-based learning strategy (Jackson Filho et al. 2016). In our group, Professors Rodolfo Vilela and José Marçal Jackson Filho have since 2010 offered the discipline of activity ergonomics in the graduate program of the FSP-USP, using the pedagogical strategy of the Conservatoire National des Arts et Métiers (CNAM), from France. Besides, Professors Rodolfo Vilela and Ildeberto Almeida (São Paulo State University/UNESP), teaching a course on accident analysis on the same graduate program, adopted a learning strategy based on the principles of Paulo Freire, which have a set of similarities with Vygotsky's theory (Alves 2012). That is to say that our pedagogical strategy in these two disciplines was already based on field data collection and discussion of the concepts of prescribed/actual work, regulation, contradictions in the work process, etc. An intervention of ergonomic work analysis (EWA) requires the construction of demand through an adequate negotiation with the applicant organizations, ethnographic fieldwork, hypothesis development, collection of systematic data, organizational diagnosis, and elaboration of a book of recommendations for solving the reformulated demand (Guérin et al. 1997).

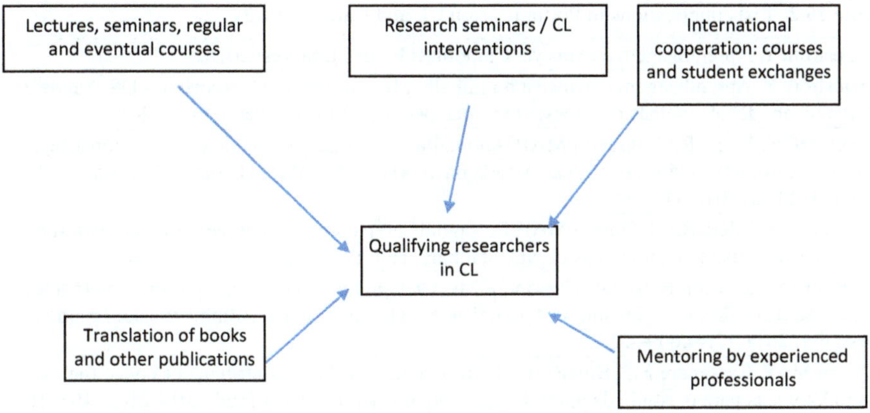

Fig. 15.2 Qualification strategies in the LM method adopted by the research group

Several students who took the ergonomics course had already manifested interest in a more global strategy of intervention that included collective and expansive learning. The foundation was ready, the environment favorable: the CL came to the group in appropriate time to further advances.

In summary, five basic strategies organized the actions for instructing students and implementing the method (see Fig. 15.2). The fifth strategy involved the exchange of professors and students from Brazilian universities, CRADLE and FIOH, as well as the participation in courses on the method conducted either in Finland or in Brazil.

Table 15.2 presents a list of the early publications on activity theory and CL in Brazil.

Another strategy was the presentation of lectures at Brazilian scientific events. The quick diffusion that resulted from this implies that there is a great demand from students and professors interested in learning the methodology, attending courses, or even integrating the research group.

Similarly, our experience in international meetings has been promising and resulted in cooperation with researchers from the CNAM, France, in a project to evaluate and improve occupational risks prevention activity in the two countries. These repercussions seem to show that CL's theoretical approach fills some gaps in occupational health and safety, in Brazil and in other countries.

The collaboration with scientists from the University of Helsinki and FIOH has resulted in fruitful exchanges between the two groups such as visits by researchers to offer courses, participation in scientific articles, the swap of experiences side-by-side and the expansion of the CRADLE site as an international space for exchange in the use of the methodology.

We were also invited to the Finnish initiative "CRADLE Global: Network of research on cultural-historical activity and formative interventions." So far, fourteen research groups in twelve countries are to integrate this network. The goal is to set up

Table 15.2 Lists publications in Portuguese on CL and learning theory

Engeström Y. Aprendizagem expansiva. Campinas: Pontes Editores; 2016.
Engeström Y. Aprendizagem expansiva no trabalho. In: Simonelli AP, Rodrigues DS. Saúde e Trabalho em debate: velhas questões, novas perspectivas. Brasília: Paralelo 15; 2013.
Lopes MGR. Vilela RAG. Querol MAP. Anomalias e contradições no processo de construção de um aeroporto: uma análise histórica baseada na Teoria da Atividade Histórico-Cultural. Cad Saúde Pública. 2018;34:1–15.
Lopes MGR, Vilela RAG, Querol MAP. Protagonismo para uma compreensão sistêmica sobre acidentes de trabalho e anomalias organizacionais. Trab Educ Saúde. 2018;2:1–26.
Querol MAP. Jackson Filho JM. Cassandre MP. Change Laboratory: uma proposta metodológica para pesquisa e desenvolvimento da Aprendizagem Organizacional. Administração: Ensino e Pesquisa. 2011;12(4):609–40.
Querol MAP. Cassandre MP. Bulgacov YL. Teoria da Atividade: contribuições conceituais e metodológicas para o estudo da aprendizagem organizacional. Gest Prod. 2014;21(2):405–16.
Silva-Macaia AA. Takahashi MAC. Vilela RAG. Querol MAP. Laboratório de Mudança: uma metodologia formativa, participativa e sistêmica para criação e transformação de sistemas de atividade. In: Toledo RF et al. (org). Pesquisa participativa em saúde: vertentes e veredas. São Paulo: Instituto de Saúde; 2018.
Silva-Macaia AA. Vilela RAG. Laboratório de Mudança: ferramenta de intervenção formativa para enfrentar o sofrimento mental e outros distúrbios relacionados ao trabalho. In: Schmidt MLG et al. Orgs. Fatores psicossociais e o processo saúde/doença no trabalho: aspectos teóricos, metodológicos, interventivos e preventivos. São Paulo: FiloCzar; 2018.
Vilela RAG et al. Expansion of the object of surveillance for occupational accidents: history and challenges underwent by a reference center aiming at prevention. Ciênc Saúde Colet. 2018;23(9):3055–66.
Virkkunen J. Newnham DS. O Laboratório de Mudança: uma ferramenta de desenvolvimento colaborativo para o trabalho e a educação. Belo Horizonte: Fabrefactum; 2015.
Virkkunen J. Vilela RAG. Querol MAP. Lopes MGR. O Laboratório de Mudança como ferramenta para transformação colaborativa de atividades de trabalho: uma entrevista com Jaakko Virkkunen. Saúde Soc. 2014;23(1):336–44.
Six expanded summaries published in the proceedings of the 18° *Congresso Brasileiro de Ergonomia*; 2016 May 23–27; Belo Horizonte, MG, Brasil. Belo Horizonte: ABERGO; 2016.
- Coluci MZO, Donatelli S, Gemma SFB, Seppänen L, Vilela RAG, Silva A, Bravo E. Preparação e negociação do Laboratório de Mudança: teoria e prática em dois casos.
- Lopes MGR, Vilela RAG, Seppänen L. O papel da dupla estimulação e da agência para entender as contradições em obras de construção de um aeroporto.
- Macaia AAS, Costa SV, Maeda ST, Querol MAP, Vilela RAG, Seppänen L. O processo de análise histórica do Laboratório de Mudanças na atividade de colaboração entre uma faculdade e um centro de saúde escola.
- Macaia AAS, Triumpho EA, Vilela RAG, Fischer FM, Seppänen L. A fase de questionamento do problema no método Laboratório de Mudanças: exemplo de uma intervenção.
- Mendes RWB, Vilela RAG, Cerveny GCO. Começar em casa: a experiência do Laboratório de Mudanças na atividade de vigilância do CEREST Piracicaba.
- Seppänen L, Takahashi MAC, Cerveny GCO, Vilela RAG. Laboratório Interfronteiras: uma ferramenta possível para a integralidade no atendimento aos acidentados de trabalho.
Entries: "Teoria da Atividade e Sistema de Atividade," "Teoria da Ação Cultural de Ações Humanas," "Método de Estimulação Dupla," "Teoria da Aprendizagem Expansiva," "Laboratório de Mudança." In: Mendes R (org). Dicionário de saúde e segurança do trabalhador: conceitos, definições, história, cultura. Novo Hamburgo: Proteção; 2018.

a web platform in order to provide support and offer a space for the exchange of experiences on the application of the interventionist methodology. On our suggestion, the network will incorporate regular seminars to discuss theoretical and practical aspects and courses about CL using the same structure we have been using in Brazil.

In addition to all the aforementioned partnerships, it is worth mentioning the importance of the collaboration within the team, which was supported by three post-doctoral students and other actors that enabled regularity and a rich work and learning.

One of the lessons learned was that the systemic transformation of work environment and its processes, while producing an organizational diagnosis, also expanded the research object. This object, a territory once exclusive of experts and scholars, who provided a diagnosis and some recommendations to local actors at the end of the investigation, now requires the construction of a shared object with the interested subjects of the organization. Co-analysis, therefore, is a precondition for the collaborative transformation.

15.3 Characteristics of Demands, Contexts, and Negotiations in the Group Experiences

Most of the interventions presented in the book took place in environments with problems related to work processes, for instance, high levels of absenteeism (CSE, school, CASA Foundation, hospital in Bauru), occupational diseases (CASA Foundation, urban cleaning), minor but frequent accidents (urban cleaning), serious work accidents (child labor), and fatal work accidents (railroad company, airport construction).

In some specific cases, such as the CSE, there was low collaboration in teaching, research and extension activities between the faculty staff and its health center, as well as the risk of interruption of healthcare provision by the health center to the general public. In the care network for injured workers, there were gaps in the available health services that did not guarantee integrality of care. The intervention in CEREST faced a context of work overload, dissatisfaction of the teams with too many obstacles to more significant actions of surveillance, as well as strained interpersonal relations.

Except for the intervention in CEREST, in all other cases the research group was actively searching to carry out field studies and to apply the CL methodology with masters or PhD students. In three cases we received a spontaneous demand for intervention from an institution that acknowledged a problem and the need for change (school, hospital, CSE). Additionally, in three cases demand came from the Labor Prosecution Office, MPT (airport, railway company, and CASA Foundation); in three others, a presentation of CL method was followed by a demand of the organization (network of attention to the injured worker, urban cleaning, CEREST);

and in one case there was a joint demand from an institution and the MPT (child labor) (see Table 15.3).

In most cases we received the demand from a health and safety supervisory body (MPT, CEREST), which made negotiation more difficult, since there was distrust on the part of the institutions/companies about how the collected information would be used. Also, in Brazilian context, any kind of negotiation for participatory research is complex. It is common in many work environments that employers and supervisors do not value workers who talk about problems in work organization.

This resistance to the interventions' implementation can be understood as the result of a limited conceptualization of the problem; otherwise, it can be due to managers not believing that there is a problem. It doesn't mean that the problem does not exist, but it may indicate that managers do not recognize it as such or are able to continue activities in spite of these problems. In other words, the activity is not in a double-bind situation (see Sect. 2.6).

We noticed that most of the institutions/companies where intervention was held were in a need state period (hospital) or double-bind state (airport, CEREST, CSE, COMETIL, railway company, public cleaning); some cases were already forming a new object with contradictions between old and new systems (Fundação CASA).

Regardless of the existence of a demand, negotiation, i.e., an initial phase of the expansive learning process, is necessary. During negotiation, the researcher needs to expand the organization's way of understanding the problem and the opinion about how it should be solved. Moreover, in this phase, researchers should support practitioners to evaluate their current attempts to develop their organization and the limitation of these attempts. Practitioners should be able to understand whether and how the CL could help the organization to overcome their current developmental challenges. The CL in intervention in the school (see Chap. 6) was noteworthy in that the lack of recognition of the problem and of adequate negotiation made it difficult for the management to expand its view of itself and intervention. One of the consequences observed in this CL was the need for the researchers-intervenors to develop strategies to solve the problems that appeared during the process, which, in the end, couldn't avoid the interruption of the school's intervention.

The intervention on the railway company shows that researchers must pay special attention to situations where the absence of demand reflects a delegitimization of the seriousness of the problem. It is common that workers' health is not seen as part of the object of the activity, and therefore, efforts to improve it are seen as a cost to be avoided. In negotiations involving fatal accidents, especially when regulatory authorities are present, a defensive behavior is typically shown by the legal team of the company. In this case, the researcher's challenge is to help management understand possible relationships between worker health and other expected outcomes of the activity.

In some cases, negotiations took more than 6 months and did not have management involvement in planning the intervention, something that may occur during an intervention (Virkkunen and Newnham 2013).

Table 15.3 Summary of cases

Chap	Case	Activity	Object	Demand origin	Main issue	Number of sessions	Methodological innovations
4	Railway company	Railway transport of passengers	Railway transport of passengers	MPT	Work accidents	Ten	Training workshops on MAPA and CL with company employees
5	CASA Foundation	Rehabilitation of law-breaking teenagers	Law-breaking teenagers	MPT	Physical, psychological, and symbolic violence	Fourteen	Introductory workshops on CL. Negotiation using double stimulation to motivate and educate managers
6	School	Basic and elementary education	Student education	SESMT	Teacher absenteeism and job dissatisfaction	Thirty (1 hour each)	
7	Hospital	Renal Replacement Therapy (TRS)	Patients with chronic renal failure for TRS	Hospital	Patients hospitalized for lack of outpatient TRS vacancies	Eight	Introductory workshop on CL
8	Child labor	To propose measures for eradicating child labor	Guaranteeing rights for families with working children and teenagers	COMETIL and MPT	Children and adolescents working with their families in the production of jewelry— -early occupational and psychosocial risk for children and teenagers	Seven	
9	Airport	Construction work on large construction projects	Airport construction	MPT	Work accidents	Six (2–3 hours)	Participation of one of the workers in the planning of the sessions
10	School health center	Education and health	Students training and healthcare users	FSP and CSE Board	Academic collaboration between FSP e CSE. Possibility of decoupling CSE from the university	Eleven	CSE users participation in every session Scenarios matrix

(continued)

Table 15.3 (continued)

Chap	Case	Activity	Object	Demand origin	Main issue	Number of sessions	Methodological innovations
11	Health network	Care network for the injured worker	Care for the injured worker	Piracicaba's CEREST	Absence of an organized flow of care for injured workers, with several assistance gaps	Thirty six	The planning group was formed by the same participants in the sessions workshops with healthcare units' professionals
12	Urban cleaning	Manual cleaning (sweeping) of public paths	Cleaning of public paths	Piracicaba's CEREST	Inadequate trash separation by the population. Changes on the sweeping map	Fourteen	Use of illustrative material and discussion cutouts
13	CEREST	Intervention in the environment and work processes in order to prevent new cases of work-related accidents/diseases	Worker's health and safety surveillance	Piracicaba's CEREST	Distribution of work. Quantity versus quality of the analysis. Isolated work	Twenty-three	Participants' training in CL methodology. Implementation of new work systems

On the Use of Double Stimulation as a Negotiation Strategy

An urban cleaning company responsible for collecting and treating urban solid waste in the city of Piracicaba, SP, presented high rates of work accidents and strain of workers, especially in the activity of manual collection of household waste. In this activity the collectors manually pick waste bags and load it into the moving trucks along the streets from house to house. The CEREST of Piracicaba, who experimented and knew the methodology of CL, mediated contact with the company's managers stimulating dialogue with the research team to apply the method, in order to develop an alternative model of activity. The negotiation began with a meeting between representatives of the company, CEREST manager and surveillance agents, and the research team. The strategy used was to show a photograph of a worker loading a truck that's making its way from house to house as a first stimulus. Before the photograph (first stimulus), we asked the representatives of the company to give their opinion about the collection model. A productive dialogue took place.

"This is not a work for people!" said the manager of the company, who coincidentally began his professional career as a waste collector and recognized the pain and the risks related to the activity.

Once there was recognition that such a model of activity is not sustainable and needed to be overcome, the team presented the CL methodology, its objectives, and conducting principles as a second stimulus, as an activity specially designed to develop and enhance activity. In a few days, the company expressed interest and openness to the development of interventionist research (see Chap. 12).

The size of the organization has an impact over demand and negotiation. Large organizations, with a very pronounced division of labor, seem to determine greater difficulty in negotiation and management participation. All in all, it becomes clear that the recognition of issue demand by practitioners is important, but it is not requisite for conducting an intervention. If there is a crisis situation, researchers can create dialogue mechanisms that facilitate the recognition of the problem during negotiation.

Example of Negotiating in an Organization with Several Departments

A railway transport company was fined by MPT due to 12 instances of workers being run over during work routines, such as maintenance on railroad tracks. The company is large with several departments. Negotiations initially involved a prosecutor from the MPT who sought the university to request support in the analysis of the accidents' causes. Its initial intent was of instruction of the investigation in order to take legal action against the company. The researchers accepted the call of the prosecutor but were more interested in implementing an in-depth study associated with organizational learning that placed greater emphasis on prevention activity than on company redress and/or punishment. The prosecutor became interested in the methodology, claiming that remedial actions usually had a low impact and left pathogenic organizational models untouched. Two meetings were held by researchers and high company staff (representative managers, legal consultants, etc.) preceded by presentations of CL methodology and of the Analysis and Prevention of Accidents' Model/MAPA (Almeida and Vilela 2010, Almeida et al. 2014). We used the same negotiation strategy with double stimulation (Engeström 2007), similar to the one presented in the case of urban cleaning. The company was summoned to a meeting with the researchers and representatives of the Ministry of Labor and Employment and MPT. The negotiation began with the presentation by the researchers of a timeline that signaled the run over accidents' dates and other disturbances such as wagon derailments and accidents involving users on a period prior to the current company's foundation (first stimulus). Next we presented the summary of the company's mission statement, which stipulates the "offer of transport with excellence, safety, and quality." The negotiation, which included high rank directors and legal advisors, resulted both in progress and difficulties: one of the directors stated that run-over accidents occurred due to inappropriate behavior (unsafe attitudes of the operators). The workers were, according to the director, "in the wrong place at the wrong time." This presented a scenario with obstacles and significant learning challenges: the complex problem of accidents reduced to human error. However, at that moment, one of the company's engineers on work safety that was at the meeting claimed that he himself, knowing the rules and aware of the risks, had already been in a near-miss runover on the tracks (multivocality). Methodology was brought in at the negotiation, but the company, still acting defensively, argued on the need for the research team to have a better understanding of the company's operation. Subsequently, it positioned itself in a way favorable to a collaborative study.

The research team acknowledged later that it failed to be fair by presenting, without sufficient prior knowledge, a negative image of the company. This reinforced the company's defensive posture, creating difficulties to cooperation, which are only being overcome with ethnographic field work, introductory workshops, and the beginning of the CL sessions.

15.4 The Process of Collecting Mirror Data and the Sessions

Once the intervention is accepted, the data collection period begins. The purpose of data collection is to enable interventionists-researchers to understand what is happening in the activity and to formulate hypotheses of the contradictions that affect it (see Sect. 2.8). The formulation of such hypotheses allows them to guide the learning process, without meaning that the interventionist is always right nor that these hypotheses are correct. It is not uncommon that these assumptions about contradictions and solutions be enriched or even changed, as was the case in the Bauru Hospital, where the initial demand was the absenteeism of the hemodialysis team and later presented itself as a situation of waiting list to perform outpatients' renal replacement therapy, focused on hemodialysis, which led to long-term hospitalization and a bigger workload for the team (see Chap. 7). Another important function of data collection is to provide mirror data to be used in the dual stimulus process during the sessions.

All the ten cases presented in this book used individual interviews in the process. In some of them, collective interviews (hospital, CSE), focus groups (CASA Foundation), and collective work analysis (Ferreira 2015) (airport, railway company, and public cleaning) were also used.

In all cases, observations of the subjects' actions within their activity system were made and, in some of them, such as the airport, the CSE, and the CASA Foundation, also participant observation. In other cases, such as that of COMETIL, it involved specific observations of meetings in which workers participated. At the school, during the sessions, the actions of local management actors were recorded on video and discussed with them in individual interviews.

In all cases except the one of urban cleaning CL, document analyses were carried out, mostly of institutional documents. In some cases it included also statistics on shifts, absenteeism, work accidents, accident or incident investigation reports, official MPT documents, and, in the CASA Foundation CL, academic publications and films about the history of the institution.

We learned that suppressing or investing too little time and resources in data collection, as in the school intervention, mainly prevents the understanding about the activity's development, necessary to plan expansive learning actions and to the dynamics of the sessions. An incomplete ethnographic work also rends trust and collaboration in the transformation process more difficult.

The video recordings of the CL sessions themselves were also used as mirror data, serving as the first stimulus in later sessions.

Examples of Mirror Data used in some Interventions

In the care network for injured workers' CL, during a session to stimulate the discussion of contradictions, mediators used videos from previous sessions as the first stimulus for participants to think about their concepts about work accidents and its care within the services network.

During the CEREST CL, interviews with workers and the CEREST's coordinator, a timeline, and board quantifying activities were used as mirror data. These data helped to assess the difficulties faced in the new model but also to perceive improvements in communication among the workers after the CL sessions.

The historical analysis in the CSE CL counted on participants for the collection of mirror data, which was mainly used to elaborate a timeline. Data was chosen in relation with the hypotheses elaborated by the group during the sessions in order to understand the critical events identified in the historical analysis.

For the COMETIL CL, a real case—an accident involving a five-year-old child who ingested acid used for welding jewelry pieces—was used as the first stimulus. The case became a turning point, helping participants in the expansive learning and engaging actors in the improvement of COMETIL, as well as in developing solutions on their own after the end of the intervention.

In the Bauru Hospital case, the group was presented with a transcription of an interview with a patient who'd been in hospital for more than one hundred days waiting for a hemodialysis vacancy as the first stimulus in the first session. It described the feeling of "being in a prison, even if there were no bars." The report had a strong impact, initially provoking defensive reactions from the medical team, who found themselves in a double-bind situation (having to provide assistance to patients without vacancies and subsequently reports of how unsustainable this situation was).

The experiences confirm the importance of ethnographic data collection of the present, past, and future developmental possibilities of the activity under intervention. This is intended not only to adequately select mirror data but also to help interventionists-researchers to formulate hypotheses about the historical contradictions, the developmental stage, the proximal developmental zone, and possible directions for development. Dialectically, an adequate elaboration of hypotheses about the contradictions and their historical origins helps to choose the best mirror data and second stimuli with greater potential to facilitate the group's expansive learning process.

Interventions' time lapse varied. Most of them were performed in seven to fourteen sessions, with an average duration of 2 hours each. The number of sessions was usually determined during negotiations. In the airport case we had to reduce the number of sessions while increasing the duration of each one, which led to the fatigue of the participants. We also believe that while it is possible to achieve results in a few sessions, the chances of modifying the activity system and implementing solutions are reduced. In other cases, as in the school CL, in the health network, and at CEREST, more than 20 sessions were held. At the school the duration of the sessions was shorter (1 hour), which limited the application of the first and second stimuli in a single session (see Table 15.3). In the health network and CEREST CLs, the extended number of sessions was due to the context of negotiation and agreements with various players involved (see Chaps. 11 and 13).

Two-hour CL session is ideal to resume last session's work, to use double stimulation strategy, and to avoid fatigue and lack of concentration of participants. In addition, we consider that the weekly frequency of the sessions meets the same goals. Sessions with 2-week intervals or more require a greater effort from researchers and participants in order to recall last session's themes.

These experiences also showed us that interventions require a considerable amount of time, which is different from each graduate program's deadlines. In none of the cases, a "complete" CL, going through all phases of expansive learning from start to finish, was applied within the graduate program term. In some cases, e.g., the CASA Foundation, negotiation and data collection were carried out within the master's course, and the sessions were continued during the doctorate.

15.5 Contradictions, Hypotheses, and Results

The experiences of interventions show that one of the important contributions of the method is to generate a deeper and systemic understanding of the historical changes that brought the system to its crisis. Throughout the CL, the participants were stimulated to conduct historical and disturbances analysis that helped the understanding of historical contradictions in and between the analyzed activity systems. A deeper understanding of the issues is per se one of the important results of the interventions. In some cases, the formulation of contradictions hypotheses was done during the sessions, using the model of the activity system and the concept of contradictions explicitly. In others, the problems and their causes were collectively analyzed with the participants, but the formulation of the contradictions was made by the researchers using theoretical concepts, as in the CSE case.

Next, we summarize the hypotheses of main contradictions that lead to undesirable results (absenteeism, occupational diseases, work accidents, difficulties in provision of services, and so on) in the different systems of activities studied.

The intervention in CEREST (see Chap. 13), the first to be implemented by the group, revealed that there was an increasing demand of inspection services that went beyond the capacities of the service. The group inspectors learned that changes were necessary to solve the problem and created new tools (MAPA), a new community (collaboration with MTE, MPT), as well as much deeper and more complex accident analyses. These changes lead to multiple contradictions in the system between the quantities of occurrences to be audited and the new objective, which demanded a more preventive work, focusing in much deeper interventions that demanded more time, energy, and a bigger team. During the intervention, the group designed two models of the activity system, one focusing on the quantity and the other on the quality, which were tested and implemented.

In the CSE case, historical changes were observed in the object of teaching, research, and extension activities of the university and its health center. The distancing of activities led the university to under-finance and lose interest in the actions developed in the center. In parallel, an increase on the demand of CSE health services was observed, which aggravated the internal issues and the relationship with the university. The intervention allowed a better understanding of their systemic causes, even though it did not make it through, at least at the time, in the implementation of the developed solutions due to a lack of participation of key agents from the senior management of the university.

The airport intervention revealed that contradictions were related mainly to historical law changes affecting construction works bidding, which did not guarantee that a project was a precondition for beginning the works. The project could be structured during its execution and resulted in lack of planning, rework, and other disturbances. Problems were aggravated by time pressure – delays and a tight deadline led to an almost simultaneous performance of activities that should be sequential. Like in the CSE case, this intervention helped participants to understand their problems' causes in a more historical, systemic way. However, due to an unexpected halt of the works and a limited participation of managers, they could not proceed to implement solutions.

In CASA Foundation, it became clear to us that, over the past decades, a new concept of social inclusion of young offenders appeared. Nevertheless, the changes necessary to CASA Foundation to adjust to this new model were only partial. New interventions were added to the youngsters routine, directed to psychological, social, and educational aspects; however, another part of the system still functioned in the old model of work, based on the principle of violence: its objective is to keep the youths isolated from society and not necessarily to rehabilitate them socially. Therefore, the work of one group interferes with the other. Researchers and participants interpreted the situation as a contradiction between two activity systems with different objectives – one directed to the socio-educational aspects and the other to the physical aspect and excessive disciplining of the youth. The intervention helped researchers to understand the historical causes of the issues. At the time this book was written, the intervention had not ended yet.

One of the most challenging interventions for us was the case to combat child labor, because of two reasons: first of all, because it dealt with an object distributed through an entire chain of production – a runaway object (Engeström 2006) – second, because it is not an object recognized as a problem; therefore, it has a low motivational power and does not engage the involvement of the business community. This situation led the intervention to focus on the activity of public services involved in the combat and prevention of child labor and, therefore, chose to support the work of the COMETIL. The intervention showed that the problem of child labor within the production chain is related to technological changes and changes of the market environment, which increased international competition and led entrepreneurs to adopt changes and strategies with negative social and environmental impacts. It also showed that, although public agents had a shared object, they worked in an uncoordinated fashion which compromised their results. The sessions helped the participants to design and implement instruments to improve the visualization of their shared object and the communication between agents.

At the elementary school, the problem reported initially was teachers' absenteeism, which caused different disturbances in its routine. The participating group redefined the problem to be analyzed as the distress of the teacher, and problems in the relationship between managers and the teachers came up.

Historical analysis of the work showed changes in the students' profiles and the need to create new instruments to mediate relations among several agents of the systems, such as students and the community (i.e., parents and guardians). Although both local management and teachers shared the same object, their actions were directed toward different scripts and uses of instruments, which lead to the observed conflicts. The CL uncovered the discomfort between local management and teachers in such a way that management interrupted the intervention.

In the case of Bauru Hospital, it was found that the disorders were related to an increase in the number of chronic renal patients that was not followed by a quantitative increase in infrastructure and personnel, resulting in a lack of vacancies. To deal with the issue, the nephrology team, having not the option to deny care claiming lack of resources, began to admit the patients, creating a new flow of care in which other sectors/teams (e.g., ward, acute renal failure team) provided care to these patients. This change generated a series of disturbances, since the sectors had neither the infrastructure nor the personnel to deal with the new object. Such changes led to contradictions between the elements of the specialized activity of other sectors and the "new object." The main result of the intervention was to allow a better understanding of the problem, which led its participants to think of new ways to provide the service and even to question the causes related to increased demand. Throughout the process, some options for change were outlined. After the CL, there was a search for new models of chronic renal care, based on successful experiences of a university hospital in the region (benchmarking) with a change in the care protocol and the end of the treatment's waiting line.

In the health network, changes occurred in the shared attention to the injured worker among several systems of activity. It was understood that although each one of the systems developed its part of the care, the relationship between them was problematic since they did not jointly achieve the necessary comprehensive care. The contradictions observed from the historical analysis reflected changes in the rules of the central management of the health network, which emphasized certain systems (such as urgency and emergency) to the detriment of the horizontality between the set of systems, especially primary care. The changes also led to "do more with less" contradictions, causing disturbances, for example, in the relationship between the subjects of the systems (health workers) and the object (integral care to other workers).

In the urban cleaning company (street sweeping activity), a contradiction was observed in the planning of the sweep areas, which prescribed areas that demanded large displacements and consequent physical stress on the workers. Other problems identified were related to the existence of conflicts and lack of cooperation and the invisibility of the work of the sweepers by the community. The mediators managed to overcome reading difficulties using pedagogical strategy and artifacts as figures for the workers to identify the problems in the activity system (AS). The intervention resulted in concrete proposals to overcome the two main contradictions. However, changes in management and financial difficulties faced by the company made it difficult to implement and test the solutions.

At the railway company, the demand followed 12 deaths of workers, result of being run over on the railroad, mainly in maintenance activities. One of the hypotheses of contradiction is that the company has little time to carry out maintenance activities, which could only be performed during intervals between trains or at night, when there would be no circulation of trains. In the face of demands for intervention to correct anomalies on the line, workers need to act on the line even with trains in circulation, at risk of accidents and serious incidents. Historically, the railway company resulted from the merger of three companies and inherited different cultures, including safety cultures. The unification effort has resulted in the formalization of training tools and the prescription of previously non-existent tasks. However, the emphasis is still on a top-down prescription that fails to capture the variations of real work. Due to negotiating difficulties and the fact that demand comes from outside the company, there is still some resistance in collaboration, especially from top management. To reduce this resistance, four training workshops were held on the CL methodology and MAPA.

15.6 Learning and Challenges in Implementing the Method

The experiences of conducting CLs show that, in all interventions, participants expanded their understanding of problems and therefore of the solutions that could solve them. Of the ten intervention cases from this book, four reached the steps of implementing a new model/reflecting on the process; three others developed solutions or examined and tested a new model, and three are still in progress, in the phases of modeling the solutions (see Fig. 15.3).

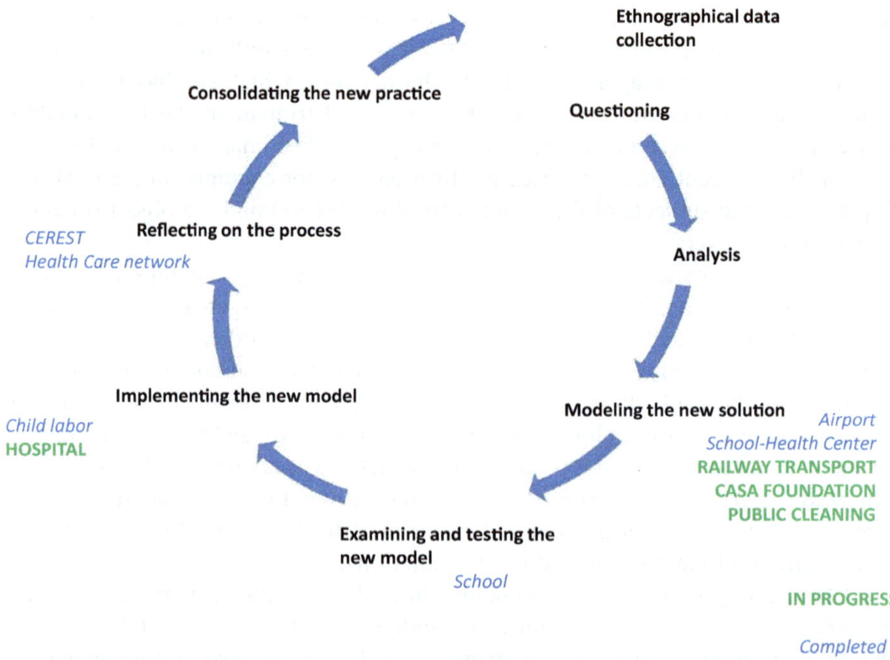

Fig. 15.3 Representation of the development stage of the ten activities under intervention in the expansive learning cycle. (Adapted from Virkkunen and Newnham 2013)

Although a systematic analysis has not been done on the creation of transformative agency in each intervention, the occurrence of expansive learning suggests agency formation in the cases. For instance, in the airport case (Lopes et al. 2018) and in COMETIL (Donatelli 2019), all types of transformative agency were observed and evaluated.

In some cases it was possible to highlight different types of agency among participants: in CEREST some people predicted new activity models, others committed themselves to concrete actions, and others carried out consequent actions to change the activity. In the hospital, the types of agencies ranged from resistance from management, criticism, explaining and envisioning new activity models, commitment, and even concrete actions (Vänninen et al. 2015).

Considering the different steps of the method, we highlight the relevance of negotiation, a fundamental aspect in the interventions since it initiates the process of expansive learning (see Chap. 2 and Sect. 15.2). The intervention cases of the railway company, urban cleaning, CASA Foundation, and school illustrate this point. This negotiation process can be improved by four strategies:

- The use of double stimulation and learning strategy since the first contact with the organization (see Chaps. 4, 5, and 12)
- Introductory workshops for better understanding of the methodology by participants (see Chaps. 4, 5, 6, and 13)
- Elaborating the intervention plan together with management (see Chaps. 4 and 5)
- In the sessions, forming mixed groups gathering researchers, management representatives, and workers to discuss progress and results of each session and plan the next ones (see Chaps. 4 and 5)

Characteristics of one of the Introductory Workshops

In face of the difficulties and challenges related to the defensive behavior at the railway company, an introductory workshop on CL and MAPA methodologies was proposed and accepted during negotiation (see Chap. 14). In total, four workshops were held and managed to achieve its objectives fully. The pedagogical strategy used on the first workshop (16 hours, 2 days in a row) was to offer a series of double stimulation so that participants reflected on the problematic situations and then mastered the theoretical concepts of the methodology.

Initially it was presented the case of an accident in which a worker was hit by a robot. The case was chosen so as to avoid defensive behavior, since it happened in a metal industry. In short, the case showed a worker operating a welding robot: he enters the robot movement region (a forbidden area). On entering, the worker was surprised by oil on the ground, slipped, and was hit by the robot's arm, in a serious accident. The metallurgical company investigated and, in line with the traditional blaming approach, concluded that from identifying the operator's error (immediate cause) and without conducting an analysis of the distal causes, the worker had committed an unsafe act. Given this first stimulus, the task assigned to four smaller groups was to re-examine the case in order to identify the gaps in this analysis and to formulate questions to better understand the systemic origins of the case. The groups worked hard and, when offered another version of the event, built by the team with the concepts of organizational analysis and MAPA, had a rich discussion which showed results diametrically opposed to the conclusions of the company: the employee had entered this prohibited area to check for defects in the products, which was impossible without that prohibited access. Therefore, it was overlooked by the boss and working team. This "swindle" was predictable because it answered to an operational need, and without it, it would be impossible to identify the problems and produce quality equipment. The reanalysis and the conceptual debate allowed participants to master basic concepts of MAPA such as prescribed and real work, change analysis, barrier analysis, bow tie model (immediate and latent causes and consequences to the event), etc. In addition, the exercise was useful for mastering the triangular model of the activity system and the contradiction concept exemplified by two irreconcilable rules: Which one to follow? The security rule, which forbids the entrance into that area, or the product quality rule, which implied approaching the equipment to check for defects in the weld?

On the second day we started with a film about Guarulhos airport, which reveals the operation of concomitant activities such as those of the airline and its check-in system, tourism company welcoming and moving passengers, safety activities, etc. The task demanded to participants was to identify the different objects in each activity and to model them using the triangular model of the activity system (Engeström 1987, 2015). With the task, participants were able to identify a network of activities with different objects and a shared object (air travel of passengers and luggage).

A second exercise was designed to make it easier for participants to identify the objects and the different subsystems that make up the railway company. The discussion was very productive and revealed the potential and contribution of the workshop by allowing participants to begin the debate about impossible tasks in the company, especially in maintenance of the permanent route activities, which is the object of investigation.

We consider the workshop our main methodological innovation. Workers that would take part of CL sessions attended it. Its main objective was to familiarize participants with the activity theory. In some cases it was performed before and in others after the data collection phase.

Another innovation that seems to have helped participants in learning during CL was their involvement in mirror data collection, as was the case with CEREST, CSE, and recently with the railway company. This strategy has led participants to make the intervention their own, gradually taking ownership of it (agency formation).

Significant practical and learning challenges were faced during the implementation processes. The first of the practical challenges identified in most of the interventions (railway company, school, CSE, airport, care network for the injured, urban cleaning, and CEREST) was the difficulty of involving management and workers, due to the traditional style of bureaucratic management where the decision making does not take into account the opinion of individuals in different hierarchical levels.

A second practical challenge was the scheduling of sessions with the participants. This difficulty occurred mainly in the CASA Foundation, COMETIL, CSE, airport, school, and hospital interventions. This challenge is probably related to the pressure for results exerted on workers, which compromises the time available for work on development actions. It may also be related to the non-recognition of the problem and consequently to a lack of motivation prior to implementing CL to make changes in the system.

Regarding session frequency and intervals, we observed that intervals longer than 2 weeks are not recommended. Although we were already conscious of it, longer spacing occurred more as a challenge of scheduling and availability of time than as an interventionist option. Therefore, we recommend that sessions' sequence and duration be considered an important point during interventions' negotiation.

A third challenge is to steer the designed solutions' implementation. Not all interventions reached the stages of designing and implementing solutions. We believe this is due to the natural slowness of the expansive learning process, which requires more time than research projects. Hence it is important to train the participants so that they continue the process after the project ends. This challenge may also be related to another one already mentioned, the lack of management involvement, which creates difficulties in implementing the solutions designed during the intervention.

These three major challenges show the importance of pre-sessions' expansive learning, even during negotiation. They demand for innovations and strategies for prior learning of managers and participants.

Experiences in Brazil show the growing importance of interventions in networks of activity systems. This type of intervention, in which there are no control centers, is more challenging to implement and requires experimentation and adaptations in CL method. In general, the unit of analysis varied from a single one to a network of activity systems. None of the activity systems analyzed, we can assure, acts in isolation but shares the object with another system or is part of a system network (Engeström 1987, 2015).

15.7 Final Remarks on the Use of CL in Occupational Health and Safety

The intervention methods usually used to resolve workplace problems are limited to specialist's diagnosis and do not involve local actors, which compromises the acceptance of the recommendations made by the interventionist. To master the CL method, disciplines, courses, and a continuous discussion seminar were implemented, to clarify the concepts, principles, and techniques of CL, as well as to share experiences and challenges. The supervision of more experienced researchers, allied to the creation of continuous opportunities for debating and reflecting, was of special relevance. These actions of supervision, training, and especially the seminars were of extreme importance for the assimilation and application of the method.

The CL method proved itself to be a favorable space for innovation, learning, and democratic experimentation of new possibilities and models of activity.

When compared to systemic approaches in the field of work health and safety, for instance, the organizational analysis of Llory and Montmayeul (2010) or constructive ergonomics (Falzon 2014), CL adds a theory of learning that's absent or insufficiently systematized in them. Seen from this angle, CL seems to be a turning point in the field of worker health and safety intervention, since it allows actors to broaden the analysis in search for contradictions that explain the system's inner anomalies (beyond technical failures or human errors—proximal antecedents), by expanding this comprehension to their historical origins (temporal expansion) and by including actors as protagonists in the establishment of diagnosis and solutions (social expansion).

These expansions were only possible thanks to the contribution of the Engeström's team, who took the propositions of Vygotsky (1978) and develop them to a higher level. The theory of cultural-historical activity and the expansive learning theory have common contact points with the critical pedagogy of Paulo Freire (1993). According to Alves (2012), Freire's and Vygotsky's approaches consider that elaborating knowledge "implies in seeing it from different angles and to see oneself reflected in it, starting from the practice and moving away from it in order to 'admire' it – to see something mediated by the dialogue with others and by the content historically produced as culture's accumulated knowledge."[1] It implies, Vygotsky would say, in a movement of de-contextualization, to leave the context while seeing oneself in it, or, in the words of Paulo Freire, "knowledge which is not related to practice does not liberate, does not generate consciousness. It is the reflective and intentional cognoscitive[2] act that produces appropriation of reality, that is, conduces to one taking upon oneself social practice and submitting it to critical analysis" (Alves 2012, p. 209–210).

In an authoritarian culture, or strongly influenced by a Taylorist approach, operators at lower hierarchical levels thinking critically can pose a threat. In these contexts, definition of work project is traditionally a task reserved to the management.

---

[1] This passage in the text brings us to the practice of confrontation recommended by the method of double stimulation that occurs in the context of the Change Laboratory.

[2] That has potential to generate knowledge.

Therefore reflecting, analyzing, and experimenting new models of activity can be seen as a threat or occupation of a territory previously exclusive to management. This primary contradiction (see Chap. 2) seems to be inherent in the capitalist mode of production and manifests itself more acutely in countries like Brazil. The increasing complexity of objects calls for a democratic division of labor and a style of management that recognize the knowledge and initiative of all hierarchical levels—that is, through extreme participatory and formative processes, therefore through power decentralization within organizations. This sharing of power, however, conflicts with the essence of power and wealth concentrating capitalism, which historically maintains vertical and essentially authoritarian, sometimes even violent, work relations and division (Metzger et al. 2012). This hypothesis of primary contradiction helps to explain the resistance of some managers who found it difficult to recognize and accept the methodology's potential. If it is true, this might be the greatest challenge for the full involvement and diffusion of the methodology, especially in private spaces or in situations that do not experience crisis.

These difficulties seem to be aggravated when there is no recognized demand or a state of crisis that drives the organization's actors to seek help. Will it be worth to insist on the implementation of CL in this scenario?

The CL seems to be more in line with a global project of expanding spaces for citizenship, peace, and sustainability, that is, a civilizational and democratic project. Thus, it seems to us that it goes against the current tendencies of globalization that intensify inter-capitalist competition, seeks to undermine the power of nation states, and encourages an "anything goes" attitude and the commodification of natural resources and human life. One can ask if the Change Laboratory has a future in a context of savage capitalism that is configured in Brazil and other continents.

These questions stand as pessimistic calls for reflection because the global scenario is also pessimistic. But we are certain that capitalism is essentially contradictory and diverse in each local, regional, and national context. By its nature the instability, the impermanence, hides under the surface of stability. Everything changes at all times. The answer to these questions may help us see in the methodology a seed to pave the way for the future, a fissure that opens within the old to make way for the new. The future, in a procedural view as the CL teaches us, needs to be cultivated within the old and thus to show that it is capable of overcoming contradictions, showing its potential to answer questions that the old cannot answer.

References

Ahonen, H., & Virkkunen, J. (2003). Shared challenge for learning: Dialogue between management and front-line workers in knowledge management. *International Journal of Information Technology and Management*, 2(1–2), 59–84. https://doi.org/10.1504/IJITM.2003.002449.

Alves, S. M. P. (2007). *Freire e Vigotski: Um diálogo entre a pedagogia freireana e a psicologia histórico-cultural* [these]. São Paulo; Universidade de São Paulo. https://doi.org/10.11606/T.48.2008.tde-16062008-133010.

Alves, S. M. (2012). *Freire e Vigotski: Um diálogo entre a pedagogia freireana e a psicologia histórico-cultural*. Chapecó: Argos.

Almeida, I.M., & Vilela, R.A.G. (2010). *Modelo de Análise e Prevenção de Acidentes de Trabalho - MAPA*. Piracicaba: CEREST.

Almeida, I.M., Vilela, R.A.G., Silva, A.J.N., & Beltran, S.L. (2014). Modelo de Análise e Prevenção de Acidentes - MAPA: Ferramenta para a vigilância em saúde do trabalhador. *Ciência & Saúde Coletiva, 19*(12), 4679–4688. https://doi.org/10.1590/1413-812320141912.12982014.

Béguin, P. (2003). Design as a mutual learning process between users and designers. *Interacting with Computers, 15*(5), 709–730. https://doi.org/10.1016/S0953-5438(03)00060-2.

Dias, A. V. C., & Lima, F. P. A. (2014). Work organization and occupational health in the contemporary capitalism. In M. G. Ribeiro (Ed.), *Frontiers in occupational health and safety* (pp. 3–31). Sharjah: Bentham Science.

Donatelli, S. (2019). *Metodologias formativas: Contribuição para o desenvolvimento colaborativo da cadeia de semijoias de Limeira [these]*. Universidade de São Paulo, São Paulo. https//:doi:10.11606/T.6.2019.tde-25042019-095410.

Edwards, A. (2007). Revealing relational work. In *Working relational in across practices: A cultural-historical approach to collaboration* (pp. 1–24). New York: Cambridge University Press.

Engeström, Y. (1987). *Learning by expanding. An activity-theoretical approach to developmental research*. Helsinki: Orienta-Konsultit.

Engeström, Y. (2006). Development, movement and agency: Breaking away into mycorrhizae activities. In K. Yamazumi (Ed.), *Building activity theory in practice: Toward the next generation* (pp. 1–43). Osaka: Center for Human Activity Theory. Kansai University.

Engeström, Y. (2007). In H. Daniels, M. Cole, & J. M. Wertsch (Eds.), *Putting Vygotsky to work: The Change Laboratory as an application of double stimulation* (pp. 363–425). Cambridge: Cambridge University Press.

Engeström, Y. (2009). The future of activity theory: A rough draft. In A. Sannino, H. Daniels, & K. Gutierrez (Eds.), *Learning and expanding with activity theory* (pp. 303–328). New York: Cambridge University Press.

Engeström, Y. (2015). *Learning by expanding: An activity-theoretical approach to developmental research*. New York: Cambridge Press.

Engeström, Y., & Sannino, A. (2010). Studies of expansive learning: Foundations, findings and future challenges. *Educational Research Review, 5*(1), 1–24. https://doi.org/10.1016/j.edurev.2009.12.002.

Engeström, Y., Virkkunen, J., Helle, M., Pihlaja, J., & Poikela, R. (1996). The Change Laboratory as a tool for transforming work. *Lifelong Learning in Europe., 1*(2), 10–17.

Falzon, P. (2014). *Constructive ergonomics*. Boca raton: CRC Press.

Ferreira, L. L. (2015). Análise coletiva do trabalho: Quer ver? Escuta. *Revista Ciências do Trabalho., 4*, 125–137.

Freire, P. (1993). *Pedagogy of the oppressed*. Texas: Continuum.

Feitosa, S. C. S. (1999). *Método Paulo Freire: Princípios e práticas de uma concepção popular de educação [dissertation]*. Universidade de São Paulo, São Paulo.

Guérin, F., Laville, A., Daniellou, F., Duraffourg, J., & Kerguelen, A. (1997). *Comprendre le travail pour le transformer – La pratique de l'ergonomie*. Lyon: ANACT.

Haapasaari, A., Engeström, Y., & Kerosuo, H. (2016). The emergence of learners' transformative agency in a Change Laboratory intervention. *Journal of Education and Work, 29*(2), 232–262. https://doi.org/10.1080/13639080.2014.900168.

Heikkila, H., & Seppänen, L. (2014). Examining developmental dialogue: The emergence of transformative agency outlines. *Critical Practice Studies, 15*(2), 5–30.

Jackson Filho, J. M., Vilela, R. A. G., Fischer, F. M., & Simonelli, A. P. (2016). Desafios pedagógicos do ensino da ergonomia da atividade na pósgraduação em saúde pública. *Ação ergonômica, 12*(2), 120–125.

Lopes, M. G. R., Vilela, R. A. G., & Seppänen, L. (2016). O papel da dupla estimulação e da agência para entender as contradições em obras de construção de um aeroporto. In *18° Congresso Brasileiro de Ergonomia; 2016 May 23–26; Belo Horizonte, Brazil*. Rio de Janeiro: ABERGO.

Lopes, M. G. R., Vilela, R. A. G., & Querol, M. A. P. (2018). Agency for a systemic comprehension of work accidents and organizational anomalies. *Trabalho, Educação e Saúde, 2*, 1–26. https://doi.org/10.1590/1981-7746-sol00128.

Llory, M., & Montmayeul, R. (2010). *L'accident et l'organisation*. Bordeaux: Préventique.

Marx, K. (2011). *Grundrisse: Manuscritos econômicos de 1857–1858: esboços da crítica da economia política*. São Paulo: Boitempo.

Metzger, J., Maugeri, S., & Benedetto-Meyer, M. (2012). Predomínio da gestão e violência simbólica. *Revista Brasileira de Saúde Ocupacional, 37*(126), 225–242. https://doi.org/10.1590/S0303-76572012000200005.

Miettinen, R. (2009). Contradictions of high-technology capitalism and the emergence of new forms of work. In A. Sannino, H. Daniels, & K. D. Gutierrez (Eds.), *Learning and expanding with activity theory* (pp. 160–175). New York: Cambridge University Press.

Scribner, S. (1997). 1925 - Vygostsky's uses of history. In E. Tobach, R. J. Falmagne, B. Parlee Mary, L. M. W. Martin, & A. S. Kapelman (Eds.), *Mind and social practice*. Select writings of Sylvia Scribner (pp. 241–265). Cambridge: Cambridge University Press.

Sannino, A. Engeström, Y., & Lemos, M. (2016) Formative interventions for expansive learning and transformative agency. *Journal of Learning Sciences, 25*(4), 599–633. http://doi.org/10.1080/10508406.2016.1204547.

Vänninen, I., Pereira-Querol, M., & Engeström, Y. (2015). Generating transformative agency among horticultural producers: An activity-theoretical approach to transforming integrated pest management. *Agricultural Systems, 139*, 38–49. https://doi.org/10.1016/j.agsy.2015.06.003.

Virkkunen, J., & Newnham, D. S. (2013). *The Change Laboratory: A tool for collaborative development of work and education*. Rotterdam: Sense Publishers.

Vygotsky, L. S. (1978). *Mind in society: The psychology of higher mental functions*. Cambridge: Harvard University Press.

Index

R. A. G. Vilela et al. (eds.), *Collaborative Development for the Prevention of Occupational Accidents and Diseases*, https://doi.org/10.1007/978-3-030-24420-0

Lightning Source UK Ltd.
Milton Keynes UK
UKHW020947100919
349515UK00002B/26/P

9 783030 244194